IUFRO Research Series

The International Union of Forestry Research Organizations (IUFRO), with its 14,000 scientists from 700 member institutions in 100 countries, is organized into nearly 300 research units that annually hold approximately 60 conferences, workshops and other meetings. The individual papers, proceedings and other material arising from these units and meetings are often published, but in a wide array of different journals and other publications. The object of the IUFRO Research Series is to offer a single, uniform outlet for high quality publications arising from major IUFRO meetings and other products of IUFRO's research units.

The editing, publishing and dissemination experience of CABI *Publishing* and the huge spread of scientific endeavours of IUFRO combine here to make information widely available that is of value to policy makers, resource managers, peer scientists and educators. The Executive Board of IUFRO forms the Editorial Advisory Board for the series and provides the monitoring and uniformity that such a high quality series requires in addition to the editorial work of conference organizers.

While adding a new body of information to the plethora currently dealing with forestry and related resources, this series seeks to provide a single, uniform forum and style that all forest scientists will turn to first as an outlet for their conference material and other products, and that the users of information will also see as a reliable and reputable source.

Although the official languages of IUFRO include English, French, German and Spanish, the majority of modern scientific papers are published in English. In this series, all books will be published in English as the main language, allowing papers occasionally to be in other languages. Guidelines for submitting and publishing material in this series are available from the Publisher, Books and Reference Works, CABI Publishing, CAB International, Wallingford, Oxon OX10 8DE, UK, and the IUFRO Secretariat, c/o Federal Forest Research Centre, Seckendorff-Gudent-Weg 8, A-1131, Vienna, Austria.

Titles Available:

1. *Forest Dynamics in Heavily Polluted Regions*
   Edited by J.L. Innes and J. Oleksyn
2. *Forest History: International Studies on Socio-economic and Forest Ecosystem Change*
   Edited by M. Agnoletti and S. Anderson
3. *Methods and Approaches in Forest History*
   Edited by M. Agnoletti and S. Anderson
4. *Air Pollution and the Forests of Developing and Rapidly Industrializing Countries*
   Edited by J.L. Innes and A.H. Haron
5. *Forests in Sustainable Mountain Development*
   Edited by M. Price and N. Butt

Methods and Approaches in Forest History

Edited by

M. Agnoletti
Department of Forestry and Environmental Sciences, University of Florence, Italy

and

S. Anderson
Forest History Society, Durham, North Carolina, USA

CABI *Publishing*
in association with
The International Union of Forestry Research Organizations (IUFRO)

CABI *Publishing* is a division of CAB *International*

CABI Publishing
CAB International
Wallingford
Oxon OX10 8DE
UK

Tel: +44 (0)1491 832111
Fax: +44 (0)1491 833508
Email: cabi@cabi.org
Web site: http://www.cabi.org

CABI Publishing
10 E 40th Street
Suite 3203
New York, NY 10016
USA

Tel: +1 212 481 7018
Fax: +1 212 686 7993
Email: cabi-nao@cabi.org

A catalogue record for this book is available from the British Library, London, UK.

Library of Congress Cataloging-in-Publication Data
Methods and approaches in forest history/edited by M. Agnoletti and S. Anderson.
p. cm. -- (IUFRO research series; 3)
Papers selected from a conference held in Florence in 1998.
Includes bibliographical references (p.).
ISBN 0-85199-420-2
1. Forests and forestry--Congresses. 2. Forest ecology--Congresses. 3. Forest management--Congresses. I. Agnoletti, Mauro. II. Anderson, S. (Steven), 1956- III. Series.

SD118.M38 2000
634.9'09--dc21 99-086187

Published in association with:

The International Union of Forestry Research Organizations (IUFRO)
c/o Federal Forest Research Centre
Seckendorff-Gudent-Weg 8
A-1131 Vienna
Austria

ISBN 0 85199 420 2

Typeset by AMA DataSet Ltd, UK.
Printed and bound in the UK at the University Press, Cambridge.

Contents

Contributors ix

Preface xiii

Introduction: the Development of Forest History Research 1
M. Agnoletti

1 Changing Roles of the Forest History Society: New Approaches to Environmental History in North America 21
S. Anderson

2 How Professional Historians Can Play a Useful Role in the Study of an Interdisciplinary Forest History 29
C. Ernst

3 Putting 'Flesh on the Carbon-Based Bones' of Forest History 35
M. Williams

4 Forest Management from Positivism to the Culture of Complexity 47
O. Ciancio and S. Nocentini

5 Economic Areas and Forest Nature: the Search for Forest Images and the Understanding of Nature in the Past 59
B. Selter

6 The Age and Size of Hazel (*Corylus avellana* L.) Stools of Nåtö Island, Åland Islands, SW Finland 67
C.-A. Hæggström

7 The Study of Charcoal-burning Sites in the Apennine Mountains of Liguria (NW Italy) as a Tool for Forest History 79
C. Montanari, P. Prono and S. Scipioni

8 Local Economic History, Environmental History and Forest History: Some Swedish Experiences and Suggestions 93
S. Gaunitz

9 Socioeconomical and Ecological Aspects of Coppice Woods History in the Lower Vosges (France) and the Black Forest (Germany) 107
R. Ostermann and A. Reif

10 The Forest History of Boreal Sweden: a Multidisciplinary Approach 119
L. Östlund and O. Zackrisson

11 Preindustrial Forests in Central Europe as Objects of Historical–Geographical Research 129
W. Schenk

12 Methods Towards Studying Historical Changes in Forest and Landscape Patterns: a Comparison between Two Nearby Socioeconomic Contexts 139
V. Robiglio

13 Temporal and Spatial Changes in a Boreal Forest Landscape: GIS Applications 157
A.-L. Axelsson

14 Comparing Damages: Italian and American Concepts of Restoration 165
M. Hall

15 Searching for Common Ground: Reconstructing Landscape History in East Africa's Eastern Arc Mountains 173
C.A. Conte

16 The 'Alnoculture' System in the Ligurian Eastern Apennines: Archive Evidence 189
S. Bertolotto and R. Cevasco

17 Between Nature and Culture: the Contribution of Anthropology to Environmental Study 203
P. Clemente, N. Breda and V. Lapiccirella-Zingari

18 Prehistoric Cultures and the Development of Woodlands 215
H. Küster

19 Integration between Genetic and Archaeobotanical Data in a Study on the Evolutionary History of *Pinus halepensis* Mill. Populations in Southern Italy 221
S. Puglisi, G. Fiorentino, R. Lovreglio and V. Leone

20 Xylology and Forest History 233
E. Corona

21 Bristlecone Pines and Tales of Change in the Great Basin 241
M.P. Cohen

22 Biodiversity and Forest Management: From Biodiversity to Geochronodiversity 251
P. Arnould, M. Hotyat and L. Simon

23 Temporal Differences in Forest History at Two Sites in Eastern North America 265
G.B. Blank

Index 275

Contributors

M. Agnoletti, Department of Forestry and Environmental Sciences, Università di Firenze, Via San Bonaventura 13, 50145 Florence, Italy

S. Anderson, Forest History Society, 701 Vickers Avenue, Durham, NC 27701, USA

P. Arnould, Center for Biogeography-Ecology, UMR 8505, CNRS/ENS, Fontenay-Saint-Cloud, France

A.-L. Axelsson, Department of Forest Vegetation Ecology, Swedish University of Agricultural Sciences, S-90183 Umeå, Sweden

S. Bertolotto, Polo Etnobotanica e Storia, Dipartimento di Storia Moderna e Contemporanea, Università di Genova, Via Balbi 6, 16126 Genova, Italy

G.B. Blank, Department of Forestry, North Carolina State University, Raleigh, NC 27695, USA

N. Breda, Università 'La Sapienza' Roma, Dipartimento di Glottoantropologia, Vicolo Sère 17/a, 31020 San Vendemiano, Treviso, Italy

R. Cevasco, Polo Etnobotanica e Storia, Dipartimento di Storia Moderna e Contemporanea, Università di Genova, Via Balbi 6, 16126 Genova, Italy

O. Ciancio, Department of Forestry and Environmental Sciences, Università di Firenze, Via San Bonaventura 13, 50145 Florence, Italy

P. Clemente, Università 'La Sapienza' Roma, Dipartimento di Glottoantropologia, c/o Dipartimento Filosofia, Via Carlo Fea 2, 00161 Rome, Italy

M.P. Cohen, Language and Literature Department, Southern Utah University, Cedar City, UT 84720, USA

C.A. Conte, Utah State University, Department of History, Logan, UT 84322, USA
E. Corona, Facoltà di Agraria – DISAFRI, Università degli Studi della Tuscia (VT), Via San Camillo de Lellis s.n.c., 01100 Viterbo, Italy
C. Ernst, Institut für Neuere Geschichte, Ludwig-Maximilians-Universität München, Trautenwolfstraße 3/I, D-80802 Munich, Germany
G. Fiorentino, ESA 5059 CNRS LAB PAAH, Institut de Botanique, 163 rue A. Broussonnet, 34095 F Montpellier Cedex 2, France
S. Gaunitz, Department of Economic History, Umeå University, S-90187 Umeå, Sweden
C.-A. Hæggström, Department of Ecology and Systematics, PO Box 7, University of Helsinki, FIN-00014, Finland
M. Hall, Institute for Environmental Studies, University of Wisconsin, Madison, WI 53706, USA; present address: Bolley Str. 1, 8006 Zurich, Switzerland
M. Hotyat, Center for Biogeography-Ecology, UMR 8505, CNRS/ENS, Fontenay-Saint-Cloud, France
H. Küster, Universität Hannover, Institut für Geobotanik, Nienburgerstraße 17, D-30167 Hannover, Germany
V. Lapiccirella-Zingari, 1 rue Argentina, 73100 Chambery, France
V. Leone, Dipartimento di Produzione Vegetale, Università della Basilicata, Via Nazario Sauro 85, 85100 Potenza, Italy
R. Lovreglio, Dipartimento di Produzione Vegetale, Università della Basilicata, Via Nazario Sauro 85, 85100 Potenza, Italy
C. Montanari, Istituto di Botanica, Università di Genova, Corso Dogali 1c, 16136 Genova, Italy
S. Nocentini, Department of Forestry and Environmental Sciences, Università di Firenze, Via San Bonaventura 13, 50145 Florence, Italy
R. Ostermann, Waldbau-Institut, Forstwiss, Fakultät der Universität Freiburg, D-79098 Freiburg, Germany
L. Östlund, Department of Vegetation Ecology, Swedish University of Agricultural Sciences, S-90183 Umeå, Sweden
P. Prono, Istituto di Botanica, Università di Genova, Corso Dogali 1c, 16136 Genova, Italy
S. Puglisi, Istituto del Germoplasma, Consiglio Nazionale delle Richerche, Via Amendola 165/A, I-70126 Bari, Italy
A. Reif, Waldbau-Institut, Forstwiss, Fakultät der Universität Freiburg, D-79098 Freiburg, Germany
V. Robiglio, Department of Forestry and Environmental Sciences, Università di Firenze, Via San Bonaventura 13, 50145 Florence, Italy
W. Schenk, Geographisches Institut der Universität Tübingen, Hölderlinstraße 12, 72074 Tübingen, Germany
S. Scipioni, Istituto di Botanica, Università di Genova, Corso Dogali 1c, 16136 Genova, Italy

B. Selter, Forstliche Dokumentationsstelle der Landesforstverwaltung, Nordrhein-Westfalen, Herbreme 2, D-59821 Arnsberg, Germany

L. Simon, Center for Biogeography-Ecology, UMR 8505, CNRS/ENS, Fontenay-Saint-Cloud, France

M. Williams, Department of Geography, University of Oxford, Mansfield Road, Oxford OX1 3TB, UK

O. Zackrisson, Department of Vegetation Ecology, Swedish University of Agricultural Sciences, S-90183 Umeå, Sweden

Preface

Most of the chapters presented in the book are based on presentations given at the International IUFRO conference 'History and Forest Resources' held in Florence (Italy) in May 1998. We would like to thank the Ente Cassa di Risparmio di Firenze and the Forest Service of the Tuscan Regional Government for sponsoring the meeting, the Italian Academy of Forestry Science for holding and promoting it, the members of the scientific and organizing committee, Florence City Council for the reception at the Palazzo della Signoria; Roberto Guicciardini for the hospitality offered at the Gargonza castle and Luigi Hermanin for leading the post-conference excursion to the Vallombrosa forest. We would also like to thank IUFRO, especially the Leader of Division 6, Niels Eler Koch and the President of IUFRO Jeff Burley for promoting the meeting and supporting the publication of this volume. We give special thanks to Elisabeth Johann, leader of the forest history subject group, for her effective collaboration and all her efforts to make the conference successful, and Lars Ostlund and John Dargavel for the collaboration and suggestions with the publication. Special thanks are also extended to the International Programs of the United States Forestry Service for support of the early editing efforts required for this volume, Gary Blank and Andrea Anderson for their assistance in editing, and Tim Hardwick and the staff of CABI *Publishing* for their assistance.

One of the most important achievements of the conference was to put together scientists from different research fields and different cultural backgrounds, stimulating an interdisciplinary approach to forest history. In order to preserve this value, the editors and CABI *Publishing* decided to allow the authors to keep their own style when reporting references. Thus, both reference styles used in scientific and historical papers are found, according to the professional background of each author. Another book based on papers from the conference (*Forest History*) has also been published by CABI *Publishing*.

M. Agnoletti
Chairman, IUFRO S6.07.02

and

S. Anderson
President, Forest History Society

Introduction: the Development of Forest History Research

M. Agnoletti

Department of Forestry and Environmental Sciences, Università di Firenze, Via San Bonaventura 13, 50145 Florence, Italy

1 Introduction

The increasing number of scholars and disciplines involved in forest history makes the debate about the approaches and methods a crucial issue for understanding the advances of the last decades. However, some consideration of methods also seems necessary to fill a gap in forest history studies which have rarely concentrated on this matter. The chapters in this volume present perspectives from anthropology, botany, ecology, economic history, genetics, geography, forestry and history, showing not only the use of different methods but also a diverse cultural approach to the matter of human/forest relationships. It is not an attempt to assess a new conceptual framework within which we can organize forest history as a method. The aim is rather to offer a broad cultural and scientific perspective for students and researchers and to focus their thinking in the current debates and findings of forest history, taking advantage of the entire field's expertise for their future work.

2 The origins of forest history

A discussion of approach and methods cannot neglect a general look at the development of forest history studies, which have a long tradition in Europe. Some attention was already given to forest history in the book of Pietro de Crescenzi *De ruralium Commodorum Libri XII*, which appeared in the 14th century, reporting the knowledge of the Romans on forestry, agriculture, hunting, fishing and gardening. However, a real interest for this subject can be seen in the 18th century in Germany, with the publication of a book by Friedrich Stisser in 1737, while lectures on this subject were held at the Forest

Academy of Tharandt some time later, in 1816–1818 (IUFRO, Sub. Gr. S6.07, 1992). During the following century more books were written, such as those by Baudrillart (1824), Behlen (1831) and Laurop (1843), but forest history was always treated together with hunting history or as an introduction to other subjects. The first important book entirely dedicated to forest history is probably the one published in Italy by Adolfo Di Berenger (1815–1895) in 1859–1863. The author describes the forest history of the Roman era and the Venetian Republic, covering everything that was considered pertinent to forestry in his times: forest utilization, technology, silviculture, management, law, botany, physiology, pathology and zoology. Also the relationships of forestry with agriculture, livestock-raising, hunting and parks are considered, together with a chapter dedicated entirely to shipbuilding. The holistic approach featured by the book shows that the 'idea' of integration already existed at the beginning of forest history studies, mostly because integration was the concept around which forestry developed in those times. But it shows also the differences between this new discipline and the developments of natural history research between the 18th and 19th century (Farber, 1982).

It is not surprising to learn that Di Berenger was born in Germany to French parents, received his forest training in Austria, but spent most of his life in Italy, becoming the first director of the Italian Forest School founded at Vallombrosa, near Florence, in 1869. This probably explains his broad cultural background, favouring the approach to a matter as complex as forest history. Starting from Roman times, Di Berenger depicts a picture of the forest environment considered essentially as a historical construct, where the former forest environment had been progressively transformed into historically and culturally determined landscapes. Knowing French and German forestry, he was also aware of the revolution occurring in the 19th century, with the widespread introduction of clear cuttings and even-aged stands to support timber production, inducing dramatic changes in many European forests. The evidence of changes in forest vegetation through time was thus given long before ecologists started to examine ecosystems with a clear consideration of temporal scale, as Clements did in 1916, but with no consideration of human influence. The reason for the early development of this approach was due to the fact that the need to clarify the origins and historical evolution is perhaps greater in forestry than in any other science. In fact, foresters deal with living organisms that have life cycles much longer than humans. Thus, the reasons for any current fact must be sought in the past, while the effects of any management choice will be seen in the future (Volz, 1998).

In Italy, Di Berenger's book did not lead to any further development of forest history studies, although some lectures on forest history were given in Vallombrosa and later in Florence, where the school was moved in the early years of the 20th century. We find some history in Cotta's lectures (1910) and references to the importance of historical investigations in forest management in the lectures of Perona (1910), but no teaching of forest history as a single discipline was conducted either in Vallombrosa or in Florence until recently.

A very different evolution affected forest history in Germany, where several books on the subject were published during the 19th century and the first decades of the 20th century. The growing interest in this discipline and the importance given to the historical approach in German forestry led to the foundation of an Institute of Forest History in the Forestry Faculty at the University of Freiburg in 1943, directed by the forester Kurt Mantel. The years after the Second World War marked important changes, with an attempt to develop a conceptual framework in forest history and the establishment of international research activity. In 1951 Von Hornstein assessed the theory and praxis of forest history, clarifying concepts such as *Forstgeschichte* and *Waldgeschichte*. He discussed the history of forestry during the first term and the history of the human–nature relationship during the second term. The latter consisted of a methodical treatment of the relationship between ecological evolution and human action. Despite the reference to ecology made by Von Hornstein, until recently most forest history studies did not consider vegetation changes.

The growing interest in natural resources in Europe was probably due also to the historical approach made by the French school of Les Annales in the 1930s and the work of economic historians, such as Bloch and Febvre, but specific historical works on forest history, such as that of Reuss (1938), were also published in France in the same period. In the USA interest in the economic role of the forest stimulated some attention to forest history, leading to the foundation of the Forest History Society in 1946, originally established as the Forest Products History Foundation, under the aegis of the Minnesota Historical Society, as reported by Anderson in Chapter 1. The big economic changes that occurred in Europe after the Second World War increased attention to landscape changes and the role of wood in economic development, and drew the attention of many historians to the environment. In Italy a history of the agrarian landscape by Sereni (1961) considered forest changes as a part of a more general discourse on the original character of the Italian landscape but tightly connected to the idea of the leading role of the city in Italian history. However, even in the first volume of a history of Italy (Sereni, 1972), as well as in the works of Braudel (1966) and Duby (1962), human society clearly remained the main subject of all the investigations. Forests were mostly considered as economic space, with little emphasis on forest changes, which were more often studied by botanists. In the Anglo-Saxon tradition, the first volume of the *Cambridge Economic History* (1941) and later books, such as that by P.W. Bamford (1956), can be considered a good example. But even in France this approach seems evident from the articles presented in one special issue of the *Revue Forestière Françaises*, entirely dedicated to forest history, published in 1962, and also in the book written by Deveze in 1965. The publication of this special issue was encouraged by the

Faculty of Literature of the University of Nancy, where historians such as Michel Deveze and M.J. Schneider carried out investigations on forest history and a course on the subject was established. In Germany foresters (surely) played a more important part in the development of forest history studies, but, looking at the works of Mantel (1990) and Hasel (1985), the interest seems more concentrated on institutional aspects, such as the history of forest laws and the history of forestry itself, while little effort was devoted to a critical analysis of human/nature relationships.

3 The development of international research

In 1963 the growing collaboration among scientists of different countries finally led to the foundation of a research group of the International Union of Forestry Research Organizations (IUFRO) dedicated to forest history. This was a clear recognition of the importance of the historical approach in forestry and the beginning of international cooperation on the subject in a way that was quite distinct from ecological studies and the work of botanists on vegetation changes. In setting up the research group, the distinction between a history of forests influenced by humans and a history of 'natural' forests not influenced by humans was recognized (Rubner, 1997). The development of ecological studies and the work of botanists implied the existence of a history of vegetation not related to human influence, which was not receiving much attention from forest historians, at least not in the German approach.

A clear view of the main trends in forest history was offered by the first international conference organized in Nancy in 1979, where forest policy, the history of forest changes and forest techniques as well as ecological aspects were discussed, showing a change in the traditional interests of forest historians (Schuler, 1980). The Nancy conference was followed by the foundation of the Groupe d'Histoire des Forêts Françaises in 1981, and in the same year new working parties were established in the IUFRO forest history group, the first of them on tropical forest history. This Working Party organized a conference on Asia, Oceania and Australasia, under the aegis of the Australian National University, and another on Latin America, under the aegis of the American Forest History Society (Steen and Tucker, 1992). The second new IUFRO group was created in 1986 and was dedicated to the history of the timber trade and timber industry, a subject already taught at the University of Forestry in Sopron (Hungary) in 1978.

The 1970s and 1980s marked a significant development in environmental studies, determining a change in the approach to forest history. One important event, affecting not only forest history studies but generally the way investigators perceived ecological changes, was probably the publication of the book *Trees and Woodland in the British Landscape* by the botanist Oliver Rackham (1976). His approach, combining woodland ecology and historical evidence, based on the understanding of traditional practice in shaping landscape, suggested that investigators use a new method, integrating traditional written sources and material evidence in order to understand forest changes. The importance of Rackham's work was soon recognized by forest historians, as demonstrated by the Nancy conference, where he presented a paper on the medieval woods in England. From a general point of view, the new achievements in ecological studies, especially in the structures and functions of forest ecosystem (Bormann and Likens, 1979), as in their dynamics, slowly moved investigations towards a growing consideration of forest changes. Nevertheless, the ecologists' awareness of the role of humans as a key factor in the evolution of forest ecosystems was still limited. In fact, with few exceptions, most of the ecological studies have always concentrated on climate and soil as the main factors affecting vegetation changes.

This situation was partly due to the general idea that most of the natural environment was not affected by anthropic pressure, as if the natural environment and humans comprised two separate areas, overlapping only in a certain amount of their space (Vogt *et al.*, 1997). More specifically, in the attempt to model ecosystem dynamics, the origin of a particular type of disturbance was, and still is, often ignored by investigators, who concentrate more on populations and reactions to disturbances. This reductionism in considering the role of humans was not shared by Eugene Odum (1959), who stated that ecology was actually an interdisciplinary science linking biological, physical and social sciences. It is, however, surprising that, even when studying landscape ecology, the role of humans was often forgotten, as in the classic text by Forman and Gordon (1986). Even some forest ecologists share this view. Recently, Piussi and Zanzi-Sulli (1997) quoted humans as only one of the factors affecting forest ecosystem dynamics when defining ecological history. Despite the little attention to humans, the development of a historical approach in ecology offered the chance to investigate long periods of time, taking into account the many factors affecting ecosystems, as rarely happens in standard investigations, which usually consider short time frames and a limited number of factors.

Closely connected to the approach of forest history is that of historical ecology, developed by anthropologists and geographers. According to C.L. Crumley (Foreword, in Balée, 1998), it seems that the American anthropologist S. Deevey first introduced this term in the early 1970s, but the most relevant studies in this subject were probably published in the 1990s. The development of ecological history took place almost independently of forest history, a discipline that was already assessing the importance of the human role as the primary factor responsible for forest changes. The assumption in historical ecology that historical and not evolutionary events are responsible for the main changes in relationships between human society and the environment (Balée, 1998) is actually shared by most forest historians. However, forest history relates these concepts only to the forest environment, while historical ecology extends this view to the whole biosphere, thus underlining an approach that differs from that of ecological history.

The greater interest in ecology slowly moved the objectives of forest history towards closer attention to the effect of human influence on forest vegetation changes, gathering the interest of a large number of scholars coming from the natural and social sciences. Greater consideration of the ecological approach was also stressed by Rubner (1986), one of the founders of the IUFRO research group on forest history. Looking at almost a dozen major conferences held on this topic, from the first in Nancy in 1979 to the one in Florence in 1998, not only the number of scientists attending, but also the disciplines and the subjects investigated, as well as the countries represented, markedly increased (Agnoletti, 1999). This growing importance of the ecological approach was officially ratified by some changes in the structure of the IUFRO group when a new working party on ecological history (originating from a project of the European Science Foundation on forest ecosystems) joined it in 1995. The presence of the new party has strongly reinforced the relations with ecology and, more generally, with the scientists involved in natural sciences, but it has also imposed a different organization of the group in order to balance all the perspectives represented. The occasion for this rearrangement was given by the Florence conference in 1998, where the former working party title 'History of the timber trade and timber industry' was changed to 'Economic and social history of the forest' (IUFRO, 1998). The new working party recognizes the relevance of many different economic and social factors affecting forest changes, not simply related to the main economic issues, but also to small-scale processes, related to local use and practices, and to cultural aspects.

As a consequence of these changes, forest history is finally embracing the whole field of social and natural sciences, encompassing humans and nature as a total phenomenon. The involvement of many scientific fields in forest history studies, as well as the confrontation among scientists with different backgrounds, has certainly benefited the advancements of studies but has also raised some problems. Although much has been done towards integrating the different approaches, there is still a separation between different scientific cultures, often termed as an opposition between 'scientists' and 'humanists' when referring to the broader aspect of environmental history. This problem was already discussed by Febvre in 1949 when he criticized the attempt to develop single histories not made by historians, isolated from the main issues of historical research. However, if this can be considered a relevant matter for the debate on the history of science in the past decades, the matter of environmental history and forest history is a very different one. In fact, we are not dealing with the historical study of one single scientific discipline, but with a problem requiring contributions from several scientific fields. As discussed by Crumley (1998), historians are often perceived as lacking an appropriate background to cope with scientific subjects, while scientists coming from the natural sciences are seen by historians as too mechanistic or reductive in presenting their models of human–environment relationships. Scientists certainly contributed to a closer understanding of forest dynamics, reducing common mistakes and misleading interpretations of forest changes and offering comparable data and methods that can be applied in different cases. However, their findings must always take into account the cultural environment where factors and processes occurred for a full understanding of their origins and their effects. Despite some cultural resistance to an integration between the two approaches, much has been done to reap the most from all new ideas and findings, no matter their origin, as long as they can contribute to the development of the discipline, which is now a very dynamic area for international research.

4 The present approaches to forest history

To some extent, the matter of approach seems more relevant than the development of new investigative techniques, as there are not that many developed expressly for forest history, but rather there are a large number of methods available due to the wide range of disciplines involved in these studies. This may lead to some substantial problems when combining the different approaches, especially considering how many scientists are now dealing with history, often with little consideration of its role, achievements and developments. In addition to recalling historical methods, the chapter by Ernst stresses that history is especially suited for describing the social and economic context in which processes are occurring and requires an appropriate perspective as well as a critical sense. As pointed out by Williams, the pluralization of forest history scholarship has perhaps neglected the fact that history cannot just recount the past. History is an intellectual process, aimed at showing things as they were, but history cannot be entirely objective as it is always influenced by cultural aspects. This is especially true in forestry, where, according to the author, 'dominant forestry discourse' had 'immense power' in interpreting the relationships between forest ecosystems and human society in many parts of the world, a view also shared by some foresters, such as Ciancio. In Ciancio *et al.*'s chapter, the authors use a classical historical approach, analysing the history of forestry and the cultural background on which forestry was based, to propose a new approach in forestry, denying the traditional mechanistic and positivist view of foresters being too concentrated on production issues, which often degraded the ecological features of forest ecosystems. This view seems to oppose a sort of historical revisionism, expressed mostly by social historians, assessing an overestimation of the negative effects of industrial needs and the claimed reduction of forests, as a general feature of human–forest relationships (Agnoletti and Anderson, 2000) – an approach probably reflecting a cultural tendency trying to reduce the idea of the 'destructiveness of humans', attenuating a general sense of guilt towards the environment, but also promoting a sort of revaluation of the role of economy.

The development and the application of industrial silviculture often provoked a crisis in many ecosystems, breaking down the relationships between local communities and the environment, causing social conflicts and affecting the cultural approach of foresters to environmental problems. This point is analysed in the chapter by Selter, where the author stresses the need for forest historians to answer questions related to the stability of former patterns of forest uses. Intentional and non-intentional human activities often had a regulating effect on the dynamic of forest ecosystems, although this cannot be considered as a standard. The need to open forest history studies to a broader view, considering social crisis and ecosystem changes connected to technical evolution, expressed by Selter, criticizes more traditional approaches, considering the main economic and legal aspects.

4.1 Local economic history

Actually, timber or fuel production is a topic quite well suited to an approach like that of economic history and is very effective in explaining not only the past but also the present history of many of the world's forest regions. As demonstrated by the conference held in Prato in 1995, economic historians may lack basic knowledge in forestry and ecology but are well acquainted with sources and historical methods. Nevertheless, we might argue that this kind of approach could be of greater help if combined with that presented in the chapters by Hæggström or Montanari *et al.*, which give valuable information on ecological changes in their study area. The feeling is that conventional historians rarely go for local-scale analyses and feel uncomfortable when considering evidence given by the changing features of forest vegetation. The limits of many historical works are often found in the lack of consideration of the effects of the historical factors on forest ecosystem dynamics. In this case, investigations remain on a very general level, often giving more information on social aspects rather than a balanced view of the interrelationships between humans and forests. There is, indeed, another way of approaching the history of natural resources, as clearly explained in the chapter by Gaunitz, dealing with local economic history as a means of understanding ecological change. In this chapter the analysis of economic history's potential for detecting environmental changes is based on small-scale investigations applied to economic problems related to the use of wood resources in an area of northern Sweden. In this way a major economic issue, such as timber production, can be placed in a well-defined forest context, offering a better chance for understanding ecological change connected to industrial development.

Scientists from other fields are more likely to look for a historical interpretation of forest changes, although their historical approach is not always correct or may lack critical analysis, as also reported in the forester's viewpoint, expressed by Ostlund *et al.* This brings us back to the problem of stimulating integration in search of common ground, an operation that must necessarily solve the matter of a mutual understanding of the problems. The possibility of understanding ecological changes and relating them to human influence is a typical skill of foresters approaching environmental history. The growing awareness among foresters of the importance of forest history is enhanced by the activity of institutions like the Forest History Society, as pointed out by Anderson, suggesting a new approach to environmental history in North America.

4.2 Techniques and practices

Techniques and practices developed to supply wood and other forest products to society have been studied since the beginning of forestry, and the evidence of past forest practices can be detected by analysing tools, techniques and the trees themselves. This view, applied to a management technique, can be clearly seen in the approach presented by Ostermann *et al.* considering coppice forest in the Alsace and Black Forests in relation to local socioeconomic development. Humans transferred culture into the forest environment, not only reducing or expanding woodlands (for many years, the favourite topic of any historian approaching forest history) but also modifying density, structure and species composition in different time or spatial scales, as a result of timber production and local practices (Agnoletti, 1998a). The development of an ecological approach in forestry has led the forester to take into account the fact that forests are a dynamic biological system, where disturbance plays an important role. Closer attention to forest ecosystem dynamics has provided an opportunity to consider the forests as 'biological archives', recording long-term changes in their features, an important achievement in the forester's approach. But, as reported by Östlund *et al.*, foresters often lack knowledge of traditional historical sources and the skills needed to combine them with field evidence.

Closer attention to sources is also stressed in the chapter by Schenk. The work of geographers has made important contributions to forest history. Concerning Germany, Schenk reports relatively scarce attention to forest history in the historical–geographical literature, but the extension of the forest in the cultural landscape of central Europe makes this an important topic for geographical research. The author opposes the possibility of establishing a subdiscipline of 'forest and wood geography' and seems disappointed by the lack of quantitative analyses, reducing the possibilities of transferring forest history information to the geographical level, and by the lack of comparative studies.

4.3 Landscape

Robiglio *et al.* and Axelsson discuss the effects of human actions in changing the features of the forests at the landscape level. An historical interpretation of landscape ecology seems of paramount importance in both ecological and forest history, as the landscape mosaic really represents one of the most important examples of how culture affects nature. The two chapters report very different situations, the first a traditionally rich historical environment in the western Italian Alps and the second a sparsely populated area of northern Sweden. Despite their differences, both areas have been heavily modified by human activities and this evidence makes historical studies an important tool for the current problems of forest management and land planning. The growing concern for the conservation of landscape resources is raising the issue of what kind of landscape we are going to preserve or restore and why, and also that the landscape represents one aspect of the cultural identity of many areas, rather than the administrative or political division. An historical approach to landscape restoration is discussed by Hall, presenting a comparison between damaged landscapes in Italy and the USA. The author shows how different cultures in the New and Old Worlds are affecting the goals and objectives of landscape restoration, and also that restoration is much older than a few years or a few decades.

The conservation of landscape resources, especially in countries with a long history of human pressure, is tightly connected with the possibility of recovering traditional forest and agricultural practices. Although these achievements are now more accepted for industrialized countries, it is interesting to see the same approach used when considering Africa's mountain forests, as Conte explains in his chapter. This study may well represent the view of an increasing number of investigators working in tropical countries (Steen and Tucker, 1992; Balée, 1998) who stress the contrast between western development patterns and local traditions, and also question widely accepted 'environmental orthodoxies' regarding African forest history. It also shows how forest history and ecological history are tightly linked, especially in the need to present policy makers with 'a clear conceptual and analytical picture' of the interrelationships between socioeconomic development and environmental changes.

4.4 Biodiversity

The matter of biodiversity is one of the most debated issues in forest ecosystem management and in many environmental studies. The historical approach proved to be very effective not only in analysing the variations in biodiversity through time, but also in explaining how it has been affected by humans. Several chapters, according to different approaches, relate to biodiversity (Axelsson *et al.*, Bertolotto *et al.*, Clemente *et al.*, Montanari *et al.*), but a very broad one is the biogeographical approach presented by Arnould *et al.*, where the author considers different kinds of diversity: specific, stational, landscape, site. Together with a reduction in biodiversity due to human action, Bertolotto *et al.* and Montanari *et al.* suggest an opposite view of the matter. The evidence that forest practices, and thus the development of certain landscape patterns, did not necessarily result in a decrease of biodiversity but often entailed an increase in the number of the species present is a recent achievement in forest history research. The practice of *Alnocoltura* described by Bertolotto *et al.* is one of the results of many investigations carried out in Liguria (Italy), under the direction of Diego Moreno *et al.* (1998). It aims at recovering ancient local practices never described in the official forestry literature. These 'local activation practices' should be preserved to suggest solutions for the management problems of the area.

4.5 The matter of scale

The importance attributed to the local scale showed by Montanari *et al.*, Bertolotto *et al.* and Conte is evident also in the anthropological studies presented by Clemente *et al.* Although the authors refer to human ecology and cultural ecology, as part of anthropologists' current interest in human–nature relations, the issue is not clearly presented as historical ecology nor does it reflect the approach of studies such as Netting's (1981). In the cases presented, the attention is not focused on the forest itself but rather on the way it is perceived by locals. The aim is to explain the global relations existing between local communities and environmental resources, presenting a different view of human action in terms of sustainable development and biodiversity. Some anthropological studies, especially in historical ecology, are still raising some points related to the long debate between biological and cultural determinism and seem to question whether greater emphasis must be placed on historical or environmental methods in approaching historical ecology (Whitehead, 1998). This is not really the point in forest history studies, where the issue is to be flexible in approach in order to detect evidence from the natural and human environment, according to the objectives of the research, assuming that humans and nature can both generate and respond to changes.

A much wider perspective in terms of time-scale is the one of palaeobotany and archaeogenetics, expressed by Küster and Puglisi *et al.* both of whom consider very long time-scales. The first chapter reports on the investigations carried out in Germany on the diffusion of beech in prehistoric times, showing that, when the first written evidence on beech forests was produced, they could not be considered as natural but had already been influenced by humans. Küster's findings are supported by other investigations suggesting a much stronger influence of humans on forest changes in prehistoric times.

Also Puglisi *et al.*, analysing some variations in populations of *Pinus halepensis* in southern Italy, suggest a possible influence of humans, starting from 5000 BC in the first case and 17,000 BC in the second. This chapter represents an important achievement of forest genetics, which has turned its attention to forest history as an important tool for explaining the variation in the genetic structure of forest tree populations.

5 Sources, techniques and methods

The matter of methods is closely related to the possibility of the various approaches to effectively detect forest changes and their relations with human society. The spread of what Rackham calls 'pseudo-histories', leading to widely accepted stories with no connection to real facts, is also related to the lack of skill in handling methods and sources, and it is much more common than one might expect. There is a feeling that, together with a more effective knowledge of environmental history, there is also a growing spread of misleading information. This is also one of the reasons why many researchers started rewriting former interpretations of forest changes, not only for Europe but also for Africa and South America (Fairhead and Leach, 1995).

The developments of forest history clearly suggest the need to master different sources, techniques and methods, but this is not a new problem in history. Marc Bloch (1953) recognized that almost any human problem needed the handling of different types of evidence but, although this may seem contradictory, the different fields of technical scholarship are distinguished by the use of different types of evidence. In forest history, very few scholars are trained to use all the different sources and methods available today, and it is also unlikely that any can master all of them. On the other hand, it seems critical for the development of this discipline to have people trained in more than one method, as in other disciplines (e.g. ecology), where recent achievements are also due to a better training of researchers when it comes to dealing with several investigative methods. From this point of view only some forest historians feel the need for such a change, so, in an attempt to find a method with which to organize natural and social scientific information, meetings and exchanges of information among different investigators are still necessary.

5.1 Written sources

Almost all of the approaches described rely on different sources (i.e. written, material and oral), for which specific investigation techniques have been developed. From this standpoint, also according to the opinion expressed by Watkins (1998), the most urgent need is to integrate information to achieve a full understanding of historical processes. This need forces investigators to find new ways in which they can choose sources, manage investigation techniques and process data, as described in this book. Each chapter presents different ways of using sources, combining different techniques and methods, according to the cultural and scientific background of each researcher and to the temporal and spatial scale of the investigations; but some of them also remind us how to use traditional sources. Concerning the most classic tool for historians, written sources, the chapter by Ernst stresses the need to assess their accuracy and their value and also suggests comparing different kinds of evidence to ascertain their reliability. In this respect, professional historians should assess the accuracy and value of historical sources, putting forest history into a broader context and investigating the micro- and macrolevels. Concerning written sources, various authors report the lack of quantitative data and long series, causing both gaps in the analysis of long-term processes affecting forest history (Schenk, and Östlund *et al.*) and a dispersion of sources.

The lack of data is often related to the fact that forests were not always a direct main interest of public administrations, while private entrepreneurs or landowners did not always keep records of activities dealing with forestry. Together with important archival collections, like those already found for France (GHFF, 1982), there are more difficulties in countries like Italy, for instance, where primary sources in the archives are often spread between a wide number of files, whose names have no direct reference to forestry matters. In this case, historical research may turn into the frustrating attempt to check an enormous quantity of files, often containing very little information. A different problem is the one related to the analysis of 'long series', also discussed in a conference organized in Germany by W. Schenk in 1998. In this case, statistical techniques are needed to process data and come to conclusions. Apart from the matter of analysing written sources, one of the needs is to transfer the information to the forest level, which is often a difficult task when not backed by a basic knowledge of forestry and ecological aspects. During the week preceding the Prato conference in 1995, a course was held for historians concerning the problem of managing forest history records. The lectures held by the author and other forest historians during that week showed that most of the difficulties of young historians were due to a lack of skill in interpreting quantitative and qualitative information on forestry and to the lack of a historical method for analysing environmental changes.

5.2 Oral sources

Different problems arise with the analysis of oral sources. These sources are normally used in anthropology and local history (Clemente *et al.*, Conte, Gaunitz) and can be invaluable in collecting reliable information on human–nature relationships in a local context, especially about techniques and practices used in forestry. There is a limited time- and spatial scale for which oral sources are suited, but even these sources must be correctly investigated; otherwise much information is lost or the conclusions drawn may result in misleading interpretations. Investigators must necessarily be trained in the object of their investigations; collaboration with ethnographers and anthropologists shows how much information can be lost when sources are not properly questioned. In addition, the common method of defining a local practice or the use of a technique from oral descriptions must always be related to a more general historical knowledge of the problems investigated, which means the possibility of having a framework in which the information can be placed and compared. The reconstruction of the relations between a local community and its wood resources, as described by Clemente *et al.* for Orgia, led to the foundation of a local forestry museum. In this case, quite typical for similar activities carried out in various regions and countries of the world, the definition of a practice and the conservation of related tools must be backed by correct information on the origin and functions of tools and techniques. Otherwise we can misunderstand the use of tools or the description of techniques made by oral sources, simply because the investigator is not aware of all the different techniques with which a procedure can be carried out or with the general technical features of the tools collected. Many of the ethnographic museums or forestry and agriculture museums, as well as archaeological museums, suffer from this problem, as they often have very large collections of material, with comparatively little information on their origin. If this is the situation in industrialized countries, we can understand the difficulties investigators face when doing research in developing countries. Books such as those by Leroi-Gourhan (1945) and Cresswell (1975) present both theoretical and field studies on traditional tools and techniques, but a clear description of the need for an objective study of techniques in environmental history is presented by Moreno in 1990. However, these researches do not focus directly on forest history and only a few studies have been made combining technical evolution and forest changes (Agnoletti, 1998a). The chapter by Ostermann and Reif clearly shows how the analyses of a silvicultural technique, implying a complex number of operations repeated for a long time in the same area, can be an effective method for understanding forest changes.

5.3 Biological archives

New opportunities in forest history have been provided by the possibility of considering forests as biological archives, but the possibility of investigating these sources has been given by a combination of new techniques and new historical methods (Östlund *et al.*). Concerning single trees, sophisticated techniques, such as dendrochronology, are now readily available for more detailed studies, yielding information on climatic events, forestry practices and forest stand evolution, as described in the chapter by Corona. The application of dendrochronology has brought about a revolution in dating archaeological evidence by overlapping ring patterns to build a master series, or chronology, and subsequently dating wood samples by comparison. But, like any other human activity, this technique too is contingent on a cultural approach (Cohen). It is one thing to establish a chronology and a very different thing to read a record from it: the way humans read tree rings and inscribe human history upon them is a cultural matter and therefore subject to interpretation. As Baillie (1982) reports in his book on tree ring dating, the closest analogy to making a tree ring chronology is a jigsaw, so that the writer of dendrochronology may be forced to distort the order in which things happened in order to impose a logical sequence of events in a particular study. Nevertheless, although a reference to the limitation of the method is necessary, dendrochronology remains one of the most original methods for collecting information, not only on climatic events but sometimes also on social changes (Baillie, 1996). Similar comments can be made on one of the most classic techniques in palaeoecology, that of pollen analysis. This technique is able to give quantitative and qualitative information on forest changes. The deposit of pollen grains on the surface of a bog, where they cannot be destroyed but are embedded in the accumulating peat, provided the possibility of detecting long-term vegetation changes in all Europe. Although these studies present fascinating pictures of forest changes in prehistoric times, there are still uncertainties as to the way pollen analyses are used to depict historical pictures, especially when the results are used to describe the history of large territories (Dubois, 1996). For this reason, it seems quite effective to support this technique with archaeological investigations, as presented by Küster in his chapter.

The way a tree was shaped to produce timber or feed livestock, its age and the presence of stumps in the territory, as well as the tree species, are all features giving precise historical information as they are related to forest or agricultural practices at the local level (Hæggström; Bertolotto *et al.*; Montanari *et al.*). In this case, transferring the investigation to a broader-scale context can offer a wider perspective on the diffusion of a single practice through different ecological, economic, social and political environments, as presented by Ostermann *et al.* for coppicing for the French and German side of the Rhine valley. Another technique used as a source of palaeoecological information, and one that is well known in archaeobotanical investigation, is the analysis of charcoal remains in former charcoal-burning sites, as described by Montanari *et al.* The anthracological spectra collected by investigating charcoal-burning sites gives reliable information on the species used in charcoal kilns, which can be related to the presence of tree species during the different periods when charcoal burning occurred. Although limited in the spatial scale of the investigation, this technique can analyse long time spans. However, the actual possibility of detecting the different layers of charcoal due to each burning event is a crucial matter for inferring the different composition of former woodlands where the wood was harvested. Also, the use of some wood species to produce charcoal was often connected to the purpose for which the charcoal was produced, so even in this case information on local practices is needed to draw conclusions. Even more sophisticated investigation techniques are those related to the analysis of the genetic structure of forest species population, made with isozyme markers, which, together with pollen analyses, can embrace a chronological range of 30,000 years. The use of these techniques is related to the availability of expensive scientific instruments, can only be done by well-trained specialists who can master the use of these techniques and must be combined with information coming from other sources.

5.4 Comparative studies

The best results from the available sources and the choice of techniques to investigate them come from the way researchers integrate information and compare their results. From this point of view, many researchers stress the need for more comparative studies, such as those presented in the chapters by Arnould *et al.*, Blank, Clemente *et al.*, Conte, Gaunitz, Ostermann *et al.*, Robiglio *et al.* and Williams. The analysis of two or more 'case-studies' seems to offer a way of comparing and integrating evidence from different areas, also using different scales. In this respect, one of the problems is selecting study areas or processes that can be effectively compared. On the other hand, different perspectives can also help us to understand the problems that researchers can encounter when applying their own method to distinct areas. In this respect, Blank proposes a fixed number of elements to take into account when comparing the history of two different sites in eastern North America, requiring a multiplicity of analytical techniques.

The effectiveness of local studies is also given by the number and quality of sources investigated. The analysis of single sources, such as written documents, although still the most popular method in forestry history studies, is diminishing in importance in favour of multiple analyses, where results from different investigation techniques are integrated in the attempt to define a more reliable picture of the processes described. One objection to this approach, pointed out by Gaunitz, is that the difference in the sources and methods of each research field makes common results difficult, but the attempts made in this direction seems to suggest a more optimistic view. Most of the investigations need to integrate their results with other methods, in order to compare their results with other findings, using the same methods that Ernst suggested for history.

The chapter by Bertolotto *et al.* presents a method where archival information is combined with field investigations focused on the historical reconstruction of 'local activation practices' and land use systems. Even more sources seem to be analysed in Robiglio *et al.*'s chapter. They use landscape ecology to analyse vegetation and land use changes, relating them to demographic evolution and economic aspects. In this study, aerial photographs, cadastres, maps and written and oral sources, as well as data from sampling plots, are integrated to study the evolution of the landscape according to human influence. The application of the historical method to landscape conservation and management is now also used in projects conducted by the Tuscan Regional Government in Italy. This project is based on local studies (Agnoletti, 1998b), where information from different sources is collected to monitor landscape evolution and to ascertain the historical and present economic value of landscape resources.

In order to facilitate the integration of different sources, especially at the geographical level, a specific tool was developed in the form of a computer program called the Geographic Information System (GIS). The integration of historical information within GIS is now a common method in forest management, and forest history studies are also conducted using this technique. The chapter by Axelsson presents the possibilities of applying this method to the evolution of biodiversity and landscape change in a boreal forest, while Arnould *et al.* present what is probably the most innovative approach to the matter of biodiversity, combining satellite imagery, radar and GIS applied to the landscape dynamics, with a procedure of 'fusion' integrating a large number of data. This method can present much information related to dynamic or static patterns, in the form of thematic maps based on quantitative and qualitative data, on variable spatial and temporal scales.

6 Conclusions

The current approaches and methods in forest history show how greatly the discipline has broadened its perspectives in recent years. Forests are shaped by humans and they also affect human society, but the possibility of detecting evidence of this mutual influence is given by our ability to combine different approaches and methods. Despite some difficulties in developing collaboration, the achievement of a better understanding of the mutual influence between humans and the biosphere is related only to the possibility of integrating environmental and cultural information, as forest history is trying to do. This task requires a better organization of international research and efforts have been made in order to rearrange the IUFRO forest history research group. In this way, the need for interdisciplinarity probably has greater chances of being fulfilled by promoting meetings and common research projects. However, much work needs to be done to make historical studies more effective. The goal is not only to be able to understand environmental changes but also to support the current issues of environmental management today, correcting the mistakes made in the past and matching the problem of integrating socioeconomic development and conservation of natural resources.

The chapters presented in this book show that forest history, as well as related disciplines, such as ecological history and historical ecology, now presents a very flexible approach, combining distinct research methods. It is important to train researchers to comprehend the complexity of this approach, without being overwhelmed by it, and to be able to decide which suggestions from each discipline will improve our ability to understand the interpenetration between humans and the forest in all temporal and spatial scales. Integrating the methods is a task that still remains to be accomplished, but the present 'state of the knowledge' already provides a consistent background to make historical investigations an effective tool in the hands of scientists and managers dealing with the complex tasks facing forest policy at the beginning of the new millennium.

References

Agnoletti, M. (1998a) *Segherie e foreste nel Trentino dal medioevo ai giorni nostri* [*Sawmills and forests in Trentino from the Middle Ages to the present*]. Museo degli usi e costume delle genti Trentine, San Michele All'Adige.

Agnoletti, M. (1998b) Landscape evolution on a central Tuscan estate between the eighteenth and twentieth centuries. In: Kirby, K.J. and Watkins, C. (eds) *The Ecological History of European Forests*. CAB International, Wallingford, UK, pp. 117–127.

Agnoletti, M. (1999) History and forest resources. *IUFRO News* 28, 1.

Agnoletti, M. and Anderson, S. (2000) *Forest History: International Studies on Socio-economic and Forest Ecosystem Change*. CAB International, Wallingford, UK.

Baillie, M.G.L. (1982) *Tree-ring Dating and Archaeology*. The University of Chicago Press, London.

Baillie, M.G.L. (1995) Dendrochronology provides an independent background for studies of the human past. In: Istituto Internazionale di Storia Economica 'F.Datini', Atti della XXVII Settimana di Studi, *L'uomo e la Foresta, secc. XIII–XVIII*, Prato, 8–13 May 1995, a cura di S. Cavaciocchi, Collana Atti delle settimane di studi ed altri convegni number 27, Florence.

Balée, W. (1998) Historical ecology: premises and postulates. In: Balée, W. (ed.) *Advances in Historical Ecology*. Columbia University Press, New York.

Bamford, P.W. (1956) *Forests and French Sea Power 1660–1789*. University of Toronto Press, Toronto.

Baudrillart, J.J. (1824) *Histoire des forêts et de leur legislation: discours préliminaire au traité générale des eaux et forêts*. Paris.

Behlen, S. (1831) *Lehrbuch der deutschen Forst und Jagdgeschichte*. Frankfurt.

Bloch, M. (1953) *The Historian's Craft*. Vintage, New York.

Bormann, F.H. and Likens, G.E. (1979) *Pattern and Process in a Forested Ecosystem*. Springer-Verlag, New York.

Braudel, F. (1966) *Le Méditerranée et le monde méditerranéen à l'époque de Philippe II*. Colin, Paris.

Cambridge Economic History (1941) Vol. I, *The Agrarian Life of the Middle Ages*. Cambridge.

Clements, F.E. (1916) *Plant Succession: an Analysis of the Development of Vegetation*. Carnegie Institute, Washington, DC.

Cotta, A. (1910) Appunti dalle lezioni di Tecnologia Forestale. Biblioteca dell'Istituto di Assestamneto Forestale, University of Florence, Florence (manuscript).

Cresswell, R. (1975) *Eléments d'ethnologie. 2. Six approches.* Colin, Paris.

Deveze, M. (1965) *La vie de la forêt française au XVIe siècle.* Jean Touzot, Paris.

Di Berenger, A. (1859–1863) *Dell'antica storia e giurisprudenza forestale in Italia.* Longo, Treviso–Venice.

Dubois, J.J. (1996) Espaces et paysages forestiers du Nord-Ouest de la France du XIII au XVIII siècles: l'apport de la biogéographie historique. In: Istituto Internazionale di Storia Economica 'F.Datini', Atti della XXVII Settimana di Studi, *L'uomo e la Foresta, secc. XIII–XVIII*, Prato, 8–13 May 1995, a cura di S. Cavaciocchi, Collana Atti delle settimane di studi ed altri convegni n 27, Florence.

Duby, G. (1962) *L'Economie rurale et la vie des campagnes dans l'Occident médiéval.* Montaigne, Paris.

Fairhead, J. and Leach, M. (1995) *Misreading the African Landscape: Society and Ecology in a Forest–Savanna Mosaic.* Cambridge University Press, Cambridge.

Farber, P.L. (1982) Research traditions in eighteenth century natural history. In: Montulenti, G. and Rossi, P. (eds) *Lazzaro Spallanzani e la Biologia del Settecento.* Olschki, pp. 397–403.

Febvre, L. (1949) *Vers un autre histoire.* Revue de métaphisique et de morale.

Forman, R.T.T. and Gordon, M. (1986) *Landscape Ecology.* John Wiley & Sons, New York.

Groupe d'Histoire des Forêts Françaises (GHFF) (1982) *Histoire des forêts françaises. Guide de recherche.* CNRS, Institut d'Histoire Moderne et Contemporaine, Copédith.

Hasel, K. (1985) *Forstgeschichte–Ein Grundgriss für Studium und Praxis.* Paul Parey, Hamburg and Berlin.

IUFRO, Res. Gr. S6.07 (1998) Forest history in Germany. *News of Forest History* 27, 3–4.

IUFRO, Sub. Gr. S6.07 (1992) Forest history in Tharandt and Eberswalde. *News of Forest History* 16/17.

Kirby, K.J. and Watkins, C. (1998) *The Ecological History of European Forests.* CAB International, Wallingford, UK.

Laurop, C.P. (1843) *Das Forst und Jagdwesen.* Stuttgart.

Leroi-Gourhan, A. (1945) *Milieu et Techniques.* Albin Michel, Paris.

Mantel, K. (1990) *Wald und Forst in der Geschichte. Ein Lehr–und Handbuch.* M. & H. Schaper, Alfed-Hannover.

Moreno, D. (1990) *Dal documento al terreno.* Il Mulino, Bologna.

Moreno, D., Cevasco, R., Bertolotto, S. and Poggi, G. (1998) Historical ecology and post-medieval management practices in alder woods (*Alnus incana* (L.) Moench) in the northern Appennines, Italy. In: Kirby, K.J. and Watkins, C. (eds) *The Ecological History of European Forests.* CAB International, Wallingford, UK.

Netting, R.McC. (1981) *Balancing on an Alp. Ecological Change and Continuity in a Swiss Mountain Community.* Cambridge University Press, London.

Odum, E. (1959) *Fundamentals of Ecology.* W.B. Saunders, Philadelphia.

Perona, V. (1910) Appunti dalle lezioni di Assestamento ed Estimo Forstale. Biblioteca dell'Istituto di Assestamento, University of Florence, Florence (manuscript).

Piussi, P. and Zanzi-Sulli, A. (1997) Selvicoltura e storia forestale. *Annali dell'Accademia Italiana di Scienze forestali* 46, 25–42.

Rackham, O. (1976) *Trees and Woodland in the British Landscape.* Dent and Sons, London.

Reuss, E. (1938) *Une légende forestière: le tire et aire. Etude historique sur l'ancien traitement des futaies feuillues, envisagé au point de vue de l'assiette des coupes*. Berger-Levrault, Nancy–Paris.
Revue Forestière Française 6, 1962.
Rubner, H. (1986) Brauchen wir eine neue Forstgeschichte? *Forstarchiv* 57(1), 29–31.
Rubner, H. (1997) Ein Menschenalter internationaler forstgeschichtlicher Forschung. *Schweiz. Z. Forstwes.* 148(2), 163–166.
Schuler, A. (1980) Wald und Zivilisation. *Schweiz. Z. Forstwes.* 131(4), 373–374.
Sereni, E. (1961) *Storia del paesaggio agrario italiano.* La Terza, Bari.
Sereni, E. (1972) Agricoltura e mondo rurale. In: *Storia d'Italia. I caratteri originari.* Einaudi, Turin.
Steen, H.K. and Tucker, R.P. (1992) *Changing Tropical Forests.* Durham.
Stisser, F. (1737) *Forst und Jagdhistorie der Deutschen.* Jena.
Vogt, K.A. *et al.* (1997) *Ecosystems.* Springer, New York.
Volz, K.-R. (1998) *Wozu Forstgeschichte – zur Bedeutung der Forstgeschichte Heute.* Forstverein für Nordrhein-Westfalen, Tagungsbericht zur Vortragsveranstaltung 'Forstgeschichte', 9–23.
Von Hornstein, F. (1951) *Wald und Mensch.* Otto Maier, Ravensburg.
Watkins, C. (1998) *European Woods and Forests.* CAB International, Wallingford, UK.
Whitehead, N.L. (1998) Ecological history and historical ecology: diachronic modelling versus historical explanation. In: Balée, W. (ed.) *Advances in Historical Ecology.* Columbia University Press, New York.

Changing Roles of the Forest History Society: New Approaches to Environmental History in North America

1

S. Anderson

Forest History Society, 701 Vickers Avenue, Durham, NC 27701, USA

In January of 1996 the Forest History Society (FHS) and the American Society of Environmental History began joint publishing of a new journal entitled simply *Environmental History*. This joint effort is reflective of new collaborative efforts needed to expand the blossoming field of environmental history. For over a decade it has been postulated the field is ripe for interdisciplinary opportunities, potentially able to bridge the gap between forestry, science and technology, social history, geography, anthropology, archaeology and ecology. The journal *Environmental History* is playing a role in coaxing interdisciplinary efforts to occur.

While the journal will continue to publish the best and most diverse scholarship, it is expected that there will be increased integration of science in environmental history; continued analysis of class, gender and race issues; continued land use and ecological history analysis; and growing opportunities for transnational research. The latter seems to parallel the forestry profession that is coming to grips with landscape-scale analysis, cross-ownership management and global marketplaces. We will need to move towards comprehensive integration of research in related subfields in order to more fully explore a world environmental history.

The journal reflects one way the FHS is committed to participating in these endeavours. The FHS will maintain its historical programme emphasis in library and archives, expand its bibliographic reference and referral services to the broader field of environmental history, and develop new programmes in education and outreach. The FHS will continue to explore and support new and productive international opportunities.

Note: Environmental History is available through membership in the FHS at US$40.00 annually. Add $8.00 for postage if outside the United States. Members also receive the *Forest History Today* magazine, the *Forest Timeline* newsletter and any new issues of the Issues Series booklets. Joint membership of FHS plus the American Society for Environmental History is also available for only US$55.00 plus $8.00 postage if international. Send cheques made payable to the Forest History Society at: 701 William Vickers Avenue, Durham, NC 27701, USA, or call 919/682-9319 for credit card orders.

1 Introduction

In January of 1996 the journal *Environmental History* was first published as a joint effort of the Forest History Society (FHS) and the American Society of Environmental History (ASEH). It replaced both *Forest and Conservation History* published by the FHS and *Environmental History Review* published by the ASEH. While this joint effort may reflect growing needs for collaboration in forest and conservation history, the foundations of the FHS remain strong and of utmost priority. This chapter provides an overview of the FHS, indicates some new areas of focus and uses it as an entry to discuss the role of forest and conservation history in the larger field of environmental history.

The FHS was established in 1946 to study forest and conservation history. The early efforts of the FHS (originally established as the Forest Products History Foundation under the Minnesota Historical Society) centred on the identification, collection and preservation of historical material. In the 1960s and 1970s the FHS added significant capabilities in interpreting and disseminating these materials, most notably marked by the *Journal of Forest History* (later *Forest and Conservation History*) and numerous book-length publications. The present mission of the FHS is to study forest and conservation history and apply it to current issues in order to affect the future and improve human welfare.

Current programmes are threefold, including: library and archives; research and publication; and education and outreach. The library and archives are the foundation of the Society. The library contains over 6500 volumes on forest and conservation history. The archives house collections from national level organizations such as the Society of American Foresters, the American Forest and Paper Association and the American Forestry Association in addition to about 60 other collections of national and international significance. The FHS also houses other unique collections including 25,000 photographs that trace forestry and logging in North America from the 1880s to the present and a news-clipping collection from the 1880s to the 1920s that spans the development of the early conservation movement in North America (this collection includes original letters from Bernard Fernow and others regarding matters of international forestry). Recently we acquired the forest history collection from the US Forest Service which also includes some international forestry topics.

For several decades the Society has been developing two unique and powerful databases, one a forest and conservation history annotated bibliography and the other an archival database which locates significant collections. Both are international in nature but are probably much more complete for North America than other parts of the world. Submissions of other bibliographies or other databases are welcome in order to make it even more complete and useful. These FHS databases are now online and fully searchable through our web pages (http://www.lib.duke.edu/forest/).

The research and publications programme focuses on historical issues important to today's decisionmakers as they plan for the future. The Society engages in comprehensive research that results in book-length publications and also produces an annual magazine entitled *Forest History Today*. The Society has an active oral history programme, having conducted over 250 oral histories with leaders and workers in forestry and conservation, a portion of which have been international in scope. Our most recent publication in this regard was *Plantation Forestry in the Amazon: the Jari Experience* that is a set of four oral histories with forestry consultants who worked with Daniel K. Ludwig on his mammoth forestry project in Brazil. The Society's academic journal, *Environmental History*, features original articles, book reviews and bibliographic and archival reports.

An awards programme is also focused on research and publications. Perhaps the opportunity of most interest to international researchers are the Alfred D. Bell Fellowships that provide travel grants for those wishing to study at the Society's library and archives. Up to US$750 is provided to individuals in North America and up to US$1500 to individuals at other international locations. Our Blegen and Hidy Awards recognize the best research in *Environmental History* and other journals, respectively. The FHS also recognizes the best journalism about forest and conservation history found in newspapers, general circulation magazines and broadcast media.

Through our education and outreach programme, the FHS engages outside audiences to improve the understanding of forest and conservation history. Although the Society does not engage in advocacy, it does seek to deepen the context within which forest policies are debated. As part of this effort, the Issues Series booklets provide short, objective, readable accounts that bring a historical context to today's most pressing natural resource issues. These booklets are used in education programmes for landowners, teachers, college students and policymakers. In addition, the Society is partnering with other institutions to develop teacher's guides and student curricula in forest and conservation history and encouraging the use of primary materials in the learning process.

In 1995, my predecessor, Harold K. Steen, led the Board of Directors of the FHS through a long-range planning process. In addition to reiterating our commitment to the library, archives and publications aspects of the programmes, the Board articulated new goals and strategies in both education (conveying an understanding of forest history to a wide variety of publics) and to actively exploring and responding to new dimensions of international forestry. This translates into the following change in operations. While traditional efforts of the FHS have been North American in focus, we will now look at our operations in the global context and take advantage of international opportunities.

The Board of Directors articulated the values of the FHS in the long-range plan. These also form the basis for our approach to collaboration between forest and conservation history and the larger field of environmental history. The FHS believes in:

1. Continuity of human events and that knowledge of the past provides insights to guide the future.
2. Forests as entire ecosystems, comprising the land, soils, water, trees and all vegetation, animals and their relationships to human life including their environmental, economic, social and cultural values.
3. Interconnectedness of forests and human events and our responsibility to be a part of an international network among those with shared interests and values.
4. Preservation of records of historical significance, and maintaining their quality and integrity to serve present and future historians.
5. Scholarship and the importance of applying intellectual skills to advance excellence in the study, analysis and interpretation of forest and conservation history.
6. Sharing information and knowledge among all with an interest in forests and conservation, and contributing to public enlightenment about the values that the forests have had for mankind.
7. Partnerships and the ensuing synergy among those who wish to support, preserve, study and apply our heritage of forest and conservation history.

Indicative of our values in scholarship and collaborative efforts, and reflective of our long-range goals in outreach and international opportunities, our current journal entitled *Environmental History* represents our new contribution to scholarly efforts. Published jointly with the ASEH, the journal is now 3 years old and enjoying a welcome reception. For the FHS, the journal *Environmental History* is the latest evolution of an effort that started in 1957 with a newsletter called *Forest History*. In 1958 the newsletter evolved into an illustrated quarterly journal and in 1974 it was renamed the *Journal of Forest History*. In 1985, the journal was billed 'the scholarly journal with the longest record of covering environmental issues' (White, 1985). In 1990, it was renamed *Forest and Conservation History*, which was replaced in 1996 with the current *Environmental History*. The new journal also has its roots in the journals *Environmental Review* and *Environmental History Review* published by ASEH.

The result of the merger was a strengthening of the journal. It not only resulted in a higher level of scholastic quality but also produced cost efficiencies and opportunities to engage environmental historians in forest history. It is a sign of the role that forest history can play in the larger field of environmental history. During its 3-year history, the journal has published articles under the broad umbrella of environmental history that explore the place of human societies within changing ecosystems; cultural concepts about humankind's relationship to nature; and environmental politics and policy. Its contributions facilitate the understanding of the reciprocal influences of nature and society with the end goal being to improve human effectiveness at living within the landscape.

If one examines the articles published in *Environmental History* during its brief 4-year history as a surrogate for the general status of the environmental history field, then some insights may present themselves regarding both the role of forest history and the emerging focus of the broader field. There are three primary areas that deserve particular attention, including: (i) the need for interdisciplinary research; (ii) the broad examination of environmental justice; and (iii) emerging transnational analysis.

2 Interdisciplinary research in environmental history

For over a decade environmental historians have been heralding the opportunity, if not the necessity, for interdisciplinary research in environmental history (White, 1985; Worster, 1988). Both White and Worster identify the obvious connections between environmental historians and historical geographers, but with increasing regularity similar notions have been purported for environmental history with the fields of anthropology (Goody, 1996), economics (O'Conner, 1997) and the history of science and technology (Worster, 1997). In *Environmental History*, two articles have touted the interdisciplinary benefits that can be accrued between environmental history and social history (Taylor, 1996; Truett, 1997). Of noted importance as well is the need to ensure the integration of more 'science' into environmental history. The science field most noted in this regard is ecology (Christensen, 1989) but could include forestry, plant pathology, entomology, botany, biology and a host of other disciplines as needed and determined by issues of concern.

While some professionals have expressed concern over the epistemological differences between the traditions behind natural science and humanities such as history, it is certainly feasible that the comparative methodology can offer a great deal to interdisciplinary research. The role, then, of foresters and forest history in environmental history is apparent. Foresters can contribute to the development of land use histories; knowledge of the previous use of technologies; silvicultural responses to land treatments; the extent of ecosystem change and other ecological data; and in policy analysis they have a vital role in the field of environmental history. The necessity of understanding the interrelationships between ecological and historical change is clear if we are to be more effective in living within the landscape.

3 Broad examination of environmental justice

There is a growing body of knowledge that examines how environmental challenges and social inequalities are connected. McGurty in her 1997 article in *Environmental History* traces the origins of the environmental justice movement, citing the limitations of the environmental movement as a root cause. In addition, evaluation of the relative effects of environmental policy decisions can help us more adequately prepare for impending resource conflicts and learn from past experiences.

A consistent stream of articles have appeared during the first three volume years of *Environmental History* that deal with language, class, gender, race, ethnicity and other demographical aspects of environmental history. For example, in 'Class and environmental history: lessons from the "War in the Adirondacks"' Jacoby (1997) describes how the establishment of the Adirondack Park was a struggle between urban and rural forces that resulted in the alienation of the local inhabitants from their surroundings. Such strands are also addressed in a variety of ways by Spence (1996) examining Native American relationships to management of National Parks in the United States; by Boone (1996) in describing flood control efforts in Montreal, Canada; by Taylor (1996) and Jacoby (1997) in the siting of toxic industries and waste dumps; and Aagesen (1998) in the history and conservation of the monkey-puzzle tree in Argentina and Chile. It is clear that continued attention to these pursuits in forest and environmental history will be made as we deal with increasing issues of sustainability in the future.

4 Transnational analysis

Because environmental issues are not confined within national boundaries; because environmental ideas and movements have been international in scope; and because local decisions can have influential impacts in the global marketplace, increasing attention is being given to transnational analysis as an approach in environmental history (Tyrell, 1991). In addition to global marketplace effects, both the consistent chasm that breaks the connections between local populations and local resources and agriculture's impact on global forests and related resources represent commonalities that provide a certain basis for such a landscape effort (Worster, 1985). The growing and potential role of transnational institutions, especially as related to the environment (Holdgate, 1995), also points the way to new opportunities in forest and environmental history.

In *Environmental History*, transnational analysis is discussed directly by Truett (1997) while re-examining resource extraction, Mexican labour and American investment; by Wirth (1996) in discussing transboundary smoke pollution from a lead and zinc smelter in British Columbia, Canada; and Aagesen (1998) who suggests that conservation of the monkey-puzzle tree may depend on transborder cooperation between Chile and Argentina. An additional opportunity for transnational environmental history was eloquently pointed out by Flader (1998) exploring the historical relationships between citizens and their governments.

In many ways, the new attention being given to transnational analysis by environmental historians parallels the forestry profession that, especially in this decade, is coming to grips with landscape-scale analysis, cross-ownership management and global marketplaces. Efforts towards ecosystem management commonly cite the need not only for landscape-scale analysis but also for multiscale analysis, from the microsite to the global landscape. Similarly, environmental historians may consider simultaneous consideration of multiple geographical scales including local, regional, national and transnational.

References

Aagesen, D.L. (1998) On the northern fringe of the South American temperate forest: the history and conservation of the monkey-puzzle tree. *Environmental History* 3, 64–85.

Boone, C.G. (1996) Language, politics and flood control nineteenth century Montreal. *Environmental History* 1(3), 70–85.

Christensen, N.L. (1989) Landscape history and ecological change. *Journal of Forest History* 33, 116–125.

Flader, S.L. (1998) Citizenry and the state in the shaping of environmental policy. *Environmental History* 3, 8–24.

Goody, J. (1996) Man and the natural world: reflections on history and anthropology. *Environment and History* 2, 255–269.

Holdgate, M.W. (1995) Pathways to sustainability: the evolving role of transnational institutions. *Environment* 37, 16–20, 38–42.

Jacoby, K. (1997) Class and environmental history: lessons from the 'war in the Adirondacks'. *Environmental History* 2, 324–342.

McGurty, E.M. (1997) From NIMBY to civil rights: the origins of the environmental justice movement. *Environmental History* 2, 301–323.

O'Conner, J. (1997) What is environmental history? Why environmental history? *Capitalism, Nature, Socialism* 8, 3–27.

Spence, M.D. (1996) Crown of the continent, backbone of the world: the American wilderness ideal and Blackfeet exclusion from Glacier National Park. *Environmental History* 1, 29–49.

Taylor, A. (1996) Unnatural inequalities: social and environmental histories. *Environmental History* 1, 6–19.

Truett, S. (1997) Neighbors by nature: rethinking region, nation, and environmental history in the US–Mexico borderlands. *Environmental History* 2, 160–178.

Tyrell, I. (1991) American exceptionalism in an age of international history. *American Historical Review* 96, 1031–1055.

White, R. (1985) American environmental history: the development of a new historical field. *Pacific Historical Review* 54, 297–335.

Wirth, J.D. (1996) The trail smelter dispute: Canadians and Americans confront transboundary pollution, 1927–41. *Environmental History* 1, 34–51.

Worster, D. (1985) World without borders: the internationalizing of environmental history. In: Bailes, K.E. (ed.) *Environmental History: Critical Issues in Comparative Perspective*. University Press of America, Lanham, Maryland, pp. 661–669.

Worster, D. (1988) Appendix: doing environmental history. In: Worster, D. (ed.) *The Ends of the Earth: Perspectives on Modern Environmental History*. Cambridge University Press, New York, pp. 289–303.

Worster, J.D. (1997) Whither environmental history. *ASEH News* 8, 1–3.

2

How Professional Historians Can Play a Useful Role in the Study of an Interdisciplinary Forest History

C. Ernst

Institut für Neuere Geschichte, Ludwig-Maximilians-Universität München, Trautenwolfstraße 3/I, D-80802 Munich, Germany

Professional historians can assess the accuracy and value of sources, put forest history in a broader historical context and investigate micro- and macroperspectives. This can lead to a more differentiated, deepened and new understanding of the past illustrated with examples covering wood scarcity, discourse and the reality of sustainability, *Verrechtlichung*, mode of charcoal supply, and the forest policy of the rural population.

As we all know, the role of the professional academic historian is the study of history as a discipline in its own right. How then can professional historians play a useful role in the study of forest history? There are three essential answers to this question. The contribution of historians consists of:

1. Assessing the accuracy and value of historical sources.
2. Putting forest history in a broader historical context.
3. Investigating the interrelationship between micro- and macrolevel history.

Each of these will be illustrated with an example from a study of the forest hills west of the River Rhine as they were in preindustrial times (Ernst, 1995, 1996, 1998) with general references to Germany too (Radkau and Schäfer, 1987; Schenk, 1996a; Selter, 1996).

 Methods and Approaches in Forest History (eds M. Agnoletti and S. Anderson)

1 Assessing the accuracy and value of historical sources

Sources are written documents that come from the past. Sources tend to lie. And even if they do not lie sources can tell only part of the story. Our ancestors wrote them from a certain perspective with a certain purpose. They perceived reality in the light of their specific view and purposes. Therefore sources – regardless of what kind they are – can reveal only aspects of historical reality (Radkau, 1996).

What methods and techniques can help to assess the accuracy and value of the information which sources give? History as an academic discipline has developed two methods to criticize sources:

1. Assessing the origins of the sources. Information about the origin of a source is needed to detect possible lies. Applying this method involves answering such questions as: Do we have the source in its entirety? Who produced the source? What was the occasion, the motive? Who wanted the information and why? Did the source intentionally conceal certain aspects? As you can easily see from these items, assessing the origins of sources is a *sine qua non* of historical work.
2. Contrast different kinds of sources. For example, historically researchers explored the wood scarcity in preindustrial Germany with the help of laws and forest descriptions only. By contrasting laws and descriptions with financial documents, maps and statistics we gain new insights into the development of forests and into the question of wood scarcity (Schenk, 1996b). In a second example, many sources formed part of a broad process of communication between the users of forests. The publication of materials on desirable forestry techniques was an element of 18th-century discourse. The debate about sustained-yield forestry started in preindustrial times and continues to this day. However, these are only words. The discourse about sustainability does not mean that sustainability was in reality pursued. Here, financial documents can reveal what contemporaries in fact did. In preindustrial time, Rhineland foresters and economists talked a lot about sustainability – and destroyed forests with cultivation methods that were not sustained yield at the same time.

Professional historians can help to criticize sources by assessing the origins of these sources and contrasting different kinds of sources (Östlund *et al.*, 1997). This is the first contribution of professional historians to interdisciplinary forest history.

2 Putting forest history in a broader historical context

Forests developed in political, legal, economical, demographical and social surroundings:

1. Political and legal surroundings. Rural populations were sometimes merely the passive subjects of laws that were intended to enforce new rules. However, this was not always the cause. To a large extent, laws codified long-held legal practices. Sovereigns wrote them, but often only formulated what was already occurring. So there the people had a hidden influence on (forest) policy and legislation. We can assess forest laws correctly only in view of general approaches of the relevant time. For early modern Germany, historians see a fundamental process in legislation, called *Verrechtlichung* (Schulze, 1996) – from 1500 onwards more and more laws regulated and codified more and more matters in more and more detailed ways. The pre-modern *Verrechtlichung* also covered forest matters (Allmann, 1989). However, the higher quantity and quality of forest laws only mirror the fundamental *Verrechtlichung*. More and better forest laws until 1800 are therefore not an extraordinary phenomenon of forest treatment. Rather they are a further example of a fundamental change that has more to do with a sophisticated administration.

2. Economical and demographical influences. Factors such as the degree of industrial concentration and the regional population density strongly influenced the prices of forest products (Setter, 1995). The way in which ironworks received their charcoal was especially important. During the 18th century, governments changed the mode of charcoal supply. At the beginning of the century they guaranteed the supply, and the ironworks did not have to pay anything for their fuelwood, or only small sums. After 1750 the rulers stopped these subsidies. Instead, the ironworks had to buy their charcoal at auction (Braun, 1991). As a result the prices increased. Changing the mode of price fixing was a political decision. When this is taken into account it can be seen that higher prices are not necessarily evidence for a shortage of fuelwood.

3. Social perspective. How the rural population acted in conflict situations of increasing administrative incursions is difficult to judge. An investigation of lawsuits can offer valuable insights into their patterns of behaviour. The rural population made a forest policy of their own. They had interests that were – generally speaking – comparable to those of the rulers. The rural population was interested in reaping the natural or financial benefits of the forests. This basic natural and financial concern strengthened the populations' interest in keeping or gaining sovereignty over their communal forests. Contrary to common belief, the way the rural populations used the forest was not, in general, lacking in wisdom or harmful to the forests, but was oriented by a rationale of their own. We can observe this – internally logical – policy on the part of the rural population in many political fields. Here again the treatment of forestry mirrors general tendencies.

So putting forest history in its broader context is the second contribution of historians.

3 Investigating the interrelationship between micro- and macrolevel history

Studying history on a microlevel involves looking very precisely at one single place and time. It is an Italian innovation in historiography which is now broadly accepted. Studying history at the macrolevel on the other hand explores general tendencies everywhere in past eras. Small-scale studies have broadened our knowledge of how general tendencies were perceived in small areas – if they had an effect there at all. Let us consider forest development in preindustrial times in this light and the kinds of issues which microlevel history may raise here:

- We cannot answer the question of wood scarcity generally. The situation differed: between region and region, from year to year, from social group to social group, and between construction timber and fuelwood.
- There is a strained relation between the worldwide discourse about sustainability and the practical measures on the scene. How did the two levels work together?
- How did the general tendency of *Verrechtlichung* work in local forests?
- Why did the rulers alter the modes of charcoal supply? Was it their intention to earn more money through higher prices?
- How did the political system influence the way the rural population formulated their forest policy and participated in the development of forests?

Professional historians can assess the accuracy and value of sources, put forest history in a broader historical context, and investigate the micro- and macroperspectives (Davis, 1990; Ginzburg, 1993; Lepetit, 1993). This can lead to a more differentiated, deepened and new understanding of the past. On the other hand it can lead to the complicating of a previously simple picture. The extent to which this more complex forest history is useful for the management and development of forest resources today may, as a result, diminish. But at least we will be acting on the basis of a more accurate historical picture.

References

Allmann, J. (1989) *Der Wald in der frühen Neuzeit. Eine mentalitäts- und sozialgeschichtliche Untersuchung am Beispiel des Pfälzer Raumes 1500–1800.* Duncker & Humblot, Berlin.

Braun, H.-J. (1991) *Das Eisenhüttenwesen des Hunsrücks.* THF, Trier.

Davis, N.Z. (1990) The shapes of social history. *Storia della Storiografia* 17, 28–34.

Ernst, C. (1995) Ein neuer Umgang mit Natur? Der Kondelwald im 18 Jahrhundert. In: Klaus F. (ed.) *Sobernheimer Gespräche III. Das Land an der Mosel – Kultur und Struktur.* Rheinland-Verlag, Koln, pp. 21–32.

Ernst, C. (1996) Norbert Franz, Waldreformen im 18 Jahrhundert. Die Anfänge der nachhaltigen Forstwirtschaft im 'Baumbusch' und im 'Kondelwald' im überrregionalen Vergleich. *Aufklärung* 9, 47–73.

Ernst, C. (1998) An ecological revolution? The 'Schlagwaldwirtschaft' in western Germany in the eighteenth and nineteenth centuries. In: Watkins, C. (ed.) *European Woods and Forests: Studies in Cultural History*. CAB International, Wallingford, UK, pp. 83–92.

Ginzburg, C. (1993) Mikro-Historie, Zwei oder drei Dinge, die ich von ihr weiß. *Historische Anthropologie* 1, 169–192.

Lepetit, B. (1993) Storia: questioni di scala. *Società e Storia* 16, 849–871.

Östlund, L., Zackrisson, O. and Axelsson, A.-L. (1997) The history and transformation of a Scandinavian boreal forest landscape since the nineteenth century. *Canadian Journal of Forest Research* 27, 1198–1206.

Radkau, J. (1996) Wood and forestry in German history: in quest of an environmental approach. *Environment and History* 2, 63–76.

Radkau, J. and Schäfer, I. (1987) *Holz. Ein Naturstoff in der Technikgeschichte*. Rowohlt, Reinbek.

Schenk, W. (1996a) *Waldnutzung, Waldzustand und regionale Entwicklung in vorindustrieller Zeit im mittleren Deutschland, Historisch-geographische Beiträge zur Erforschung von Kulturlandschaften in Mainfranken und Nordhessen*. Steiner, Stuttgart.

Schenk, W. (1996b) Forest development types in central Germany in pre-industrial times. A contribution by historical geography to the solution of a forest history research argument about the 'wood scarcity' in the eighteenth century. Istituto Internazionale di Storia Economica 'F. Datini' Prato (Hg.), L'uomo e la foresta secc. XIII–XVIII. Le Monnier, Firenze, pp. 201–223.

Schulze, W. (1996) *Einführung in die Neuere Geschichte*. UTB, Stuttgart.

Selter, B. (1995) *Waldnutzung und ländliche Gesellschaft. Landwirtschaftlicher 'Nährwald' und neue Holzökonomie im Sauerland des 18 und 19 Jahrhunderts*. Schöningh, Paderborn.

Selter, B. (1996) Forstgeschichte und Umweltgeschichte in Westfalen: Definitionen und Konzepte, Forschungsstand und Aufgaben. *Westfälische Forschungen* 46, 547–603.

3 Putting 'Flesh on the Carbon-Based Bones' of Forest History

M. Williams

Department of Geography, University of Oxford, Mansfield Road, Oxford OX1 3TB, UK

The writing of forest history is affected by dominant discourses that exclude other interpretations. These arose because of the way early forestry was organized and written up by active practitioners, and because of the universal tendency to see a deep division between nature and culture. It is contended that hegemonic discourses have affected management techniques, with direct consequences for the extent, composition and variety of forests. They have also led to a misreading of indigenous forest landscapes. The post-modernist approach since the 1980s has emphasized a more holistic approach and old norms have been questioned. Case studies from the Blue Mountains of eastern Oregon and Washington, the Amazonian lowlands and the Kissidougou region of Guinea in West Africa illustrate the complexity and richness of the new interpretations.

1 Forestry as a discourse

In recent years the writing of forest history has undergone significant shifts in interpretation and treatment to create a new discourse in place of the old. The term 'discourse' is used here in Foucauldian sense (Foucault, 1972, 1980) to mean a totalizing and powerful explanation that can exclude other interpretations. A dozen or so of these dominant paradigms which have dominated the history of forestry at various times in the past are outlined and discussed in Cianco and Nocentini (1997); but most of these orthodox discourses have been questioned. Writing in 1992 on the 'new forestry' in the United States, McQuillan suggests that the shift does not lie simply with the expression of ideas as put down on the page, that is to say the style and writing of forest history, but is more basic. It resides in our very attitude, understanding and knowledge of what forestry is. This ultimately affects our written expositions, and even our actions. He argues that forestry:

> . . . is first and foremost a *discourse*, a language-based expression of relationships between forests and foresters, between forests and the public at large, and, not least, between the public and its forests. It is the unfolding of this discourse that puts flesh on the carbon-based bones of forest history. (McQuillan, 1993)

Nowhere in the natural world are these nature–culture relationships more evident than in the forest, which is a part of nature that is alive, growing and consequently, over long timespans, is open to adaptation, mutation and change. The dynamic nature of the forests means that they are not perfectly balanced systems. Ecologists have long detected different levels of resistance to disturbance, i.e. forests can undergo change and return to a similar state, though not necessarily to exactly the same state. The forest does not have inherent stability, and disturbance processes often do not lead to self-regulation (Botkin, 1990). For decades, ecologists have been arguing that this is so, but few forest historians have understood it, and cling, instead, to outmoded concepts of balance, stability and equilibrium.

What matters more than ecological adjustment, however, are human actions, and these might depend on policies and paradigmatic practices, which are reflected in the prevalent forestry discourse of any time. As Shands (1993) put it succinctly in an address to foresters in the United States:

> The forests on which you practice silviculture today are the legacy of centuries of human decisions and actions. These actions were as diverse as seasonal burning by Native Americans, reforestation by the Civilian Conservation Corps, control of wildfire, and the implementation of laws, policies and programs such as the Weeks Law and the Multiple-Use Sustained Yield Act (MUSY).

While McQuillan's argument on dominant discourses relates specifically to forestry in the temperate, developed world, the same argument can be advanced for the tropical, less developed world. In this case, however, for 'foresters' substitute 'indigenous peoples' and for 'forestry' substitute 'traditional agricultural systems'. The outcome, however, is much the same; the power of the prevailing paradigm has excluded alternative explanations and actions.

2 Orthodox views of forestry paradigms

Let us start with what might be called the traditional or orthodox 'western' treatment of forests. For a long time the paradigmatic view of forestry in the past in America and the ex-British empire territories (Canada, Australia,

New Zealand, India) lay in the wide acceptance of the 'seminal role of foresters in the history of natural resource conversion and the economic pre-eminence of timber until the second half of this century' (Legg, 1988). Timber production and its conservation, together with conservation and efficiency in the use of other natural features, such as water for irrigation and power, forage for grazing, and the soil for agriculture, were the dominant 'environmental' concerns of the age (Hays, 1959; Williams, 1999). These concerns were bolstered by shared themes of expansion, prosperity and common foes.

In forestry this singular and influential 'orthodox' view was achieved because it was foresters who interpreted and publicized their own history. In the United States it was Fernow, Pinchot and Greeley; in British India it was Brandis, Von Ribbentrop and Stebbing; in Australia it was Carron; in New Zealand it was Poole; and in the United Kingdom it was Forestry Commission officials. Their accounts are a combination of pride in the tremendous achievements of a profession made defensive and 'insular by much bureaucratic infighting, political antagonism, and public apathy' during the late 19th and early 20th centuries and an almost evangelical zeal for forest conservation (Legg, 1988). They emphasized themes of the bureaucratization and professionalization of foresters, the lobbying of politicians and the wooing of the public about the need for conservation in the face of a perceived timber famine, a scarcity perhaps manufactured by them to make them more indispensable (Williams, 1989). In particular they promoted widespread and active management of the forest environment through the twin, positive themes of sustained yield and multiple-use management techniques, to a sceptical, apathetic and at times, even hostile, public. There was the concentration on the scientific method to the exclusion of local knowledge and practice, and the scientific 'expert' reigned supreme (Tobey, 1971). The era was characterized by campaigns *against* fire, insect infestation, land-grabbing monopolies, waste and even agricultural expansion, but *for* timber physics, greater efficiency and a workable mix between state and private operations, and cooperation characterized the approach and the era.

This approach presupposes seeing nature and culture as divisible – that there is a dichotomy between humans and nature. The Cartesian heritage of a distinction between mind and matter is deeply ingrained; humans and trees are regarded as binary and mutually opposed categories. Boundary problems are interesting but do not undermine the dominant conceptual schema. The forest is the stage: humans are the actors that do things on it, and to it. Thus, the clearing, management, harvesting and settlement of former forests, and the capitalistic production of its commodities all presuppose a world composed of discrete natural, inanimate objects that are appropriated, owned and alienated by animate individuals. If we add to all this what seems like the innate western attitude to 'improve' everything in nature by making it more efficient, more productive and less wasteful, and thereby try to eliminate death and decay, then the forest can be seen as a theatre for human action and change.

This approach entailed the elimination of the natural canopy forest and its replacement with uniform stands of vigorously growing young trees that produced high annual increments of wood fibre, which also meant a reduction of biodiversity. Old-growth forests, or 'ancient forests', which were only preserved in national park and wilderness areas as monuments to the past, had no value or redeeming features for the production-oriented forester. They were 'over mature' and 'decadent' and consequently harvested. But even the ideal of sustained yield was dropped eventually. It did not equate with a sustained stock of trees. In fact they were 'polar opposites . . . [as] . . . high level perpetual sustained yield is obtained only when old growth was at last converted to vigorous and usually homogeneous stands of young timber' (McQuillan, 1993).

Nowhere has this binary view been more pronounced than in the United States where a common, almost universal, environmental position is to see 'nature as virginal "otherness", an object of deification, an Eden from which sinful humans must be barred' (McQuillan, 1993). It can only be preserved by being cordoned off into national parks and wilderness areas so that wilderness is perceived as being/becoming 'an area unchanged and uninhabited by human beings' (Rosenburg, 1994; Proctor, 1998). Compared with America, the binary opposites of nature and culture have been less strongly developed in Europe. There the cumulative effect of human alteration over millennia has been so pervasive and so great that the celebrated palaeoecologist, Faegri, said recently that even in Scandinavia, often regarded as a wilderness outpost, a virgin landscape has been 'a fiction' since the Neolithic. Rather it was a 'cultural landscape' in which 'with some small and doubtful exceptions all vegetation types were created or modified by man' (Faegri, 1988).

3 A paradigm shift

But the dominant paradigm in forest history shifted during the late 1970s and 1980s which, among other things, saw a general resurgence of interest in the past and a general concern for the environment. There were, suggests Legg (1988), three major changes. First, a proliferation of interpretations. To take one example – the events in the Americas and Australia where the celebrations (or commiserations, depending on your point of view) of the 500th anniversary of the Columbian landfall in 1492 and the 200th anniversary of the entry of the British First Fleet into Sydney harbour in 1788 were fiercely contested. These events, and their impact on vegetation and indigenous peoples, led to much introspection and an increasing awareness that history could write, interpret and reinterpret in radically different ways (Sale, 1990; Denevan, 1992).

Secondly, there was a pluralization of scholarship about forest history which was clearly influenced by the multiplicity of new interpretations.

It was characterized by the general growth of historical scholarship into the development of forest policy and forest conservation, into corporations and prominent individuals, into forest transport and machinery, labour and unions. In addition, the close connection between forests and conservation, national parks and general environmental issues led to a vast expansion of concerns and much rewriting of forest history. The forest industry has claimed that it was the first real environmental movement; that it is in the forefront of conservation and stewardship; and that sustained yield and multipurpose management are environmentally friendly. Critics would reply that forestry has had too cosy a relationship with big business – in fact that forest-based industries are the epitome of big business, and have become the unacceptable face of giant transnational corporations. For ecocentricists nothing in nature should be destroyed. Past policies are seen as misguided and based on a dangerous mythology, particularly that of sustained yield. Many of these criticisms have been taken on board.

To a certain extent both of these trends are reflected in the way in which the title of the main publication of the Forest History Society has changed. When founded in 1957 it was called the *Forest History Newsletter*; which became the journal *Forest History* (1959); then the *Journal of Forest History* (1975); and then *Forest and Conservation History* (1989). Recently, in 1996, it merged with the journal of the American Society for Environmental History, and became *Environmental History*.

Finally in this trilogy of broad changes, there is the polarization of interpretations about forest history, which I want to look at in more detail. Legg instances a number of themes under this heading but I want to develop one that relates back to my opening remarks about forestry as a discourse.

4 The nature of history and the 'new forestry'

As the example of the Columbian sesquicentenary shows, the past is open to many interpretations. This is not a new idea. The British historian J.H. Plumb, in his *Death of the Past* (1969), made a distinction between 'the past' and 'history'. The past can be used in a variety of ways, he said, to:

> ... explain the origins and purpose of human life, to sanctify institutions of government, to give validity to class structure, to provide moral example, to vivify cultural and educational process, to interpret the future, to invest the individual human life or a nation's with a sense of destiny ...

amongst many others. But, as he points out, the 'past' is never 'history', although parts of it may be historical:

> History, like science, is an intellectual process. Like science too, it requires imagination, creativity and empathy as well as observation as accurate as a scholar can make it.

History, then, is an intellectual process we engage in order to see things as they were. Therefore history cannot be entirely objective or uncontaminated by prevailing attitudes and values, which are all part of the social context. There have long been different ideological and philosophical stances as well as modes of social theorizing in the conception and practice of history, and it is a fallacy to suppose that the reconstruction of the past, whether in forests, families or factories, can be thoroughly objective and freed from the values of the enquirer.

In particular, the 'new forestry' had taken on board the language and position of post-modernism. There is a requirement to see all the familiar things in new terms, reshuffling the canonical feelings and values. The use of post-modernist language is typical, with words like 'diversity', 'eclecticism', 'plurality', 'multifaceted', 'multiple', 'contested' and 'power' being typical.

Like much post-modernist writing it is easier to say what the 'new forestry' *is not* than what *it is*, and there is a strong sense of departure from old norms and explanations but little sense of destination. None the less, we can say that the new discourse on the forest and about forestry is not about binary opposites of nature and culture. Like any other form of land management, the profession has to engage in the uncertain process of redefining its relationships with nature, and what is thought of as natural (e.g. trees, competition). Secondly, we can also say that it is not the old Prussian model of a highly regulated forest in order to achieve high levels of production (Lowood, 1990). Concern for regulation is being curbed and concerns for biodiversity, ecological system complexity, aesthetics, protection of natural fauna and flora, clean air and water, amongst others, are being substituted. Production has not been abandoned, but it has certainly been played down in the light of economic prudence. Thus, by 1993 Alan McQuillan could say that the 'old forestry' concerns in the United States had undergone 'traumatic transformation' and given way to the 'new forestry'. This meant a broadening of values away from fibre production, efficiency and even conservation, to include aesthetics and sustainability of natural ecological processes, and an innovative recognition of the complexity of the subject.

But a little thought makes one realize that it is not that simple. The relationship between humans and the environment is a two-way, reflexive process, in which human knowledge, practices and theory often 'produce' that natural environment, which then becomes a human artefact and human resource, which in turn affects human behaviour. In William Cronon's words, we have to be aware of the way 'our expectations and ideas of the natural world shape the way we use and manipulate it' (Cronon, 1995a, b).

The reality of the new discourse can be brought out in a number of examples from North America, tropical Latin America and Africa.

5 Forest dreams, forest nightmares

Nancy Langston's study of the forests of the Blue Mountains of eastern Oregon and Washington, aptly entitled *Forest Dreams, Forest Nightmares* (Langston, 1995) illustrates well the meaning of the new discourse. In the 19th century white Americans came to the Blue Mountains and exploited the natural resources of the rivers, forests and grass that they found there. The *Pinus ponderosa* pine was brutally and indiscriminately stripped out, non-farmable cut-over covered large areas, and fires ravaged the rest. Vast numbers of cattle and then later sheep, grazed out the natural pastures, and wildlife was hunted and exterminated.

By 1906 the professional federal forest managers took over in order to save the Blue Mountains, and applied their 'orthodox' discourse with the forest. The forests were withdrawn from the public domain and placed in the new forest reserve system. With the best possible motive they set out to stem the corruption of land disposal, and the greed of speculation and resource misuse. Their job was to achieve Pinchot's conservation aim of 'the greatest good of the greatest number for the longest time' in the face of a perceived timber famine. With selfless, enlightened scientific knowledge they imposed a management system designed to maximize the yield of the forest and restore the grasslands of the mountains. Less productive old-growth timber (decadent and immature) was cut out and replaced by younger, faster-growing trees, that ensured a supply of timber and protected the watersheds. Buoyed up by a sense of urgency and optimism they saw themselves as '. . . the heroes in an almost epic struggle against big business, outsiders, insects, disease – against the forest itself' (Langston, 1995).

But by so doing the Forest Service had turned the forest dream into a forest nightmare, and by about 1980 the results of past practices were becoming evident. Light logging, with shorter and shorter cutting cycles of 30 years or less, smashed the ecology. The interests of fire management liquidated the ponderosa pines, but introduced firs. Insect infections became more frequent and more devastating, and wildlife collapsed.

At the heart of the problem was the age-old American propensity of seeing 'nature and people as separate' (Langston, 1995). The Forest Service has moved in a direction of substantially protecting ecological integrity and biological diversity. Its adoption of ecosystem management in 1992 represented the adoption of a new discourse, one that includes among other things a focus on sustainable systems rather than commodity timber crops. It also encompasses a new focus on natural system characteristics such as death, decay and waste, as well as inefficiency, uncertainty and redundancy. While this may be an improvement over the previous commodity focus it is still an attempt to control natural processes. In abandoning one discourse, the Forest Service has adopted another.

6 Misreading the landscape

In the developing world, the break with the paradigms of the past has come less with the positive act of promoting tree growth (forestry) and much more with understanding the negative act of destroying trees (deforestation). Much of the tropical forest has been badly 'misread' or misunderstood, and its inhabitants unjustly accused of either apathy or wanton destruction. In a recent work on indigenous knowledge and the sociocultural aspects of trees in non-European societies, Seeland rejects the nature–culture dichotomy and stresses the idea that 'nature is culture'. He continues:

> . . . plants, animals, and human beings constitute an integrated whole, it is this specific culture that makes sense by ordering natural and social phenomena into a structured world . . . [of] . . . natural phenomena and man-made achievements, and represents time-bound selections of experience and notions which are amalgamations of nature and culture. (Seeland, 1997)

Consequently, the tropical rainforest can no longer be regarded as unscathed by human manipulation. It is not a pristine wilderness in which the indigenous inhabits were either incapable or unwilling to change anything. For example, in tropical Latin America there is widespread evidence of charcoal in the south and central Amazon, so much so that Uhl would go so far as to say that 'in much of the Amazon, it is difficult to find soils that are not studded with charcoal'. It is possible, of course, that some of this is a result of natural lightning fires, but more likely it is the result of clearing for swiddens and the past manipulation of the forest for useful plants (Uhl *et al.*, 1990). Indeed, ethnobotanists like Posey (1985), Balée (1993) and Roosevelt (1989) suggest that much of the Amazonian forest is a 'cultural artefact' or 'anthropogenic' as native peoples have developed successive resource management strategies to cope with fluctuations in population dynamics and disturbances. Thus, it is a mosaic of different ages, compositions and structures made all the more complex by the propagation of useful tree crops like nuts, palms and bamboo, and an increase in the diversity of plants. Posey (1985) concludes that even in areas where Indians have disappeared 'the hand of human manipulation and management may still be evident'. Similar arguments can be made for the Maya lowlands and other parts of tropical central America (Gomez-Pompa *et al.*, 1987). It is estimated that at present up to 40% of the tropical forest is secondary forest resulting from clearing, and that the remainder has suffered from some sort of modification.

If human manipulation of the forest by native peoples is acknowledged, the dominant discourse has been that it has been mismanaged. In particular, the imperial and colonial political overlordship of the past was convinced of the inferiority of native practices which were consequently repressed, and there was a strong conviction that the application of western development and its science and organization was superior. However, that conventional wisdom has always been challenged by social anthropologists who have shown that indigenous practice is frequently an intelligent and intelligible accommodation with environment that could well be emulated in contemporary efforts to utilize the tropical forests (Strathern, 1980; Ellen and Fukui, 1996).

Nowhere has this been more amply demonstrated than in the grassy savanna–forest mosaic of western Africa. The century-old wisdom has been that the landscape has been degraded and that the remaining patches of prominent forest – the 'forest islands' – are relicts in a sea of savanna of a formerly extensive forest cover. Ever since first French occupation of the Kissidougou landscapes of Guinea in 1893 and even into post-independence, administrators have been convinced that 'savannaization' has been the result of repeated firing of the forest by the local inhabitants. So strong was this interpretation that outsiders like foresters, politicians, colonial civil servants and modern-day aid agencies sought to take resource control away from the local inhabitants, impose repressive policies, and even criminalized certain forms of land use. In the 1970s the setting of bush fires carried the death penalty. The dominant discourse was an exercise in power which excluded other interpretations.

But Fairhead and Leach in their *Misreading the African Landscape* (1995) argue that the Kissidougou landscape is one that is filling with forests, not emptying. They present irrefutable evidence that the forests are gaining and growing in size. They also show how the local inhabitants have an intimate relationship with the forest and develop patches of forest around their settlements. Everyday activities, like cattle keeping, thatch collection, gardening, defecation, burning, and the like, contribute to the forest island development; and the new woods are important in the daily lives as sources of timber, fuelwood, food and medicines. The system has been compromised by inappropriate outside interventions, which in the environmentally conscious 1990s cynically see this degrading environment as the way to gain money, justify budgets and keep environmental institutions solvent by attracting major international funding for rehabilitation.

The history of the conventional, orthodox discourse is fascinating. The age-old concern about the relationship between tree felling, diminishing rainfall and ultimate desiccation conditioned thinking. During the 1930s there was also the conviction that the Sahara desert was growing and on the march in the northern portions of West African countries, just as the dustbowl and its dust storms seemed to be engulfing the inland plains of the United States. Amongst many researchers writing at the time, the work of Aubreville (1938), who had worked in West Africa during the 1930s, was immensely influential. The publication of his article in 1947 on 'The disappearance of the tropical forests of Africa' seemed to provide conclusive proof that science had seen the enemy, and that it was the Malthusian pressure on resources that was causing forest loss.

The overwhelming and overriding discourse of degradation has had a series of detrimental effects on local life. First, local people have been impoverished by the imposition of taxes and fines on forest use, reducing their opportunities to benefit from their resources, and diverting funds from more pressing concerns. Secondly, locals have been accused of wanton destruction and their everyday activities criminalized, and have been denied the technical validity of their ecological knowledge and development. Thirdly, they have been denied the value and credibility of their cultural forms, expressions, and even basis of morality, so that at times even their intelligence and consciousness has been denied; and fourthly, a gulf has been created between urban and rural. The perspective of the dominantly urban-based post-colonial administrators towards the rural locals has been antagonistic. They are seen as incapable of responding to modernity, reinforcing ethnic stereotypes and differentiation. In later work, Fairhead and Leach (1988) have extended their conclusions to a much wider area of West Africa.

7 Conclusion

Enough has been said, perhaps, to suggest that the dominant forestry discourse has immense power, not only to describe and interpret the landscape of trees, but also to affect the way the trees are managed, appear and respond in their natural state. In much of the western world, forests are revered and cordoned off, and those forests which have been used have been exploited as fibre factories. In the example of the Blue Mountains, the landscape has been misused, and the rewriting of the restoration by means of traditional forestry has impoverished the region. In the example of western Africa not only has the landscape been misread, but the rewriting of it in terms of the degradation discourse has denied the local people their own history. Everywhere, there is a need to appreciate a more pluralistic ecology that recognizes non-equilibrium issues.

In all such cases, it is the dominant discourse about people and trees, and the failure to see them in a more holistic way that has caused problems. Ciancio and Nocentini (1997) suggest that the traditional scientific paradigm which emphasizes rationality, analytical approaches, reductionism and a linear view should be replaced by a 'new' scientific paradigm that emphasizes intuition, synthesis, holism and a non-linear view. Simplistic reductionism must be replaced by an appreciation of complexity. It is only in this way, in the words of McQuillan, that we can put back 'the flesh on the carbon-based bones of forest history'.

Acknowledgement

The author wishes to thank Steven Anderson for some useful suggestions.

References

Aubreville, A.M.A. (1938) La forêt coloniale: les forêts de l'Afrique Occidentale Française. *Annales d'Academie des Sciences Coloniales* (Societe d'Editions Geographiques, Maritimes et Coloniales), 9, 1–245.

Aubreville, A.M.A. (1947) The disappearance of the tropical forests of Africa. *Unasylva* 1, 5–11.

Balée, W. (1993) Indigenous transformation of Amazonian forests: an example from Maranharo, Brazil. *L'Homme* 126/128, 231–254.

Botkin, D.B. (1990) *Discordant Harmonies: a New Ecology for the Twenty-first Century.* Oxford University Press, New York.

Ciancio, O. and Nocentini, S. (1997) The forest and man: the evolution of forestry thought from modern humanism to the culture of complexity, systematic silviculture and management on a natural basis. In: Ciancio, O. (ed.) *The Forest and Man.* Accademia Italiana di Scienze Forestali, Florence, pp. 21–114.

Cronon, W. (1995a) Foreword. In: Langston, N. *Forest Dreams, Forest Nightmares.* University of Washington Press, Seattle.

Cronon, W. (1995b) The trouble with wilderness: or, getting back to the wrong nature. In: Cronon, W. (ed.) *Uncommon Ground: Towards Reinventing Nature.* W. W. Norton, New York, pp. 69–90.

Denevan, W.M. (1992) The pristine myth: the landscape of the Americas in 1492. *Annals of the Association of American Geographers* 82, 369–385.

Ellen, R. and Fukui, K. (1996) *Redefining Nature: Ecology, Culture and Domestication.* Berg, Oxford.

Faegri, K. (1988) Preface. In: Birks, H.H., Birks, H.J.B., Kaland, P.E. and Moe, D. (eds) *The Cultural Landscape – Past, Present and Future*, Vols 1–2. Cambridge University Press, Cambridge.

Fairhead, J. and Leach, M. (1995) *Misreading the African Landscape: Society and Ecology in a Forest–Savanna Mosaic.* Cambridge University Press, Cambridge.

Fairhead, J. and Leach, M. (1998) *Reframing Deforestation: Global Analysis and Local Realities: Studies in West Africa.* Routledge, London.

Foucault, M. (1972) *The Archaeology of Knowledge.* Tavistock Press, London.

Foucault, M. (1980) *Power/Knowledge: Selected Interviews and Other Writings, 1972–1977.* Harvester Press, Brighton.

Gomez-Pompa, A., Flores, J.S. and Sosa, V. (1987) The 'Pet kot:' a man-made tropical forest of the Maya. *Interciencia* 12, 10–15.

Hays, S.P. (1959) *Conservation and the Gospel of Efficiency: the Progressive Conservation Movement, 1890–1920.* Harvard University Press, Cambridge, Massachusetts.

Langston, N. (1995) *Forest Dreams, Forest Nightmares: the Paradox of Old Growth in the Inland West.* University of Washington Press, Seattle.

Legg, S. (1988) Re-writing the history of forestry? Changing perceptions of forest management in the New World. In: Frawley, K.J. and Semple, N.M. (eds) *Australia's Ever Changing Forests: Proceedings of the First National Conference on Australian Forest History, Canberra, 1988.* Department of Geography and Oceanography, Canberra, pp. 223–336.

Lowood, H.E. (1990) The calculating forester: quantification, cameral science and the emergence of scientific forestry in Germany. In: Frangsmyr, T., Heilbron, J.H. and

Ryder, R.C. (eds) *The Quantifying Spirit in the Eighteenth Century.* University of California Press, Berkeley and Los Angeles, pp. 315–342.

McQuillan, A.G. (1993) Cabbages and kings: the ethics and aesthetics of new forestry. *Environmental Values* 2, 191–221.

Plumb, J.H. (1969) *The Death of the Past.* Macmillan, London.

Posey, D. (1985) Indigenous management of tropical forest ecosystems: the case of the Kayapo Indians of the Brazilian Amazon. *Agroforestry Systems* 3, 139–158.

Proctor, J.D. (1998) The social construction of nature: relativist accusations, pragmatist and critical responses. *Annals of the Association of American Geographers* 88, 352–376.

Roosevelt, A. (1989) Resource management in Amazonia before the conquest: beyond ethnographic projection. In: Posey, D. and Balée, W. (eds) *Resource Management in Amazonia: Indigenous and Folk Strategies.* Advances in Economic Botany No. 7, New York Botanical Gardens, New York, pp. 30–62.

Rosenberg, K.A. (1994) *Wilderness Preservation: a Reference Handbook.* ABC-Clio, Santa Barbara, California.

Sale, K. (1990) *The Conquest of Paradise: Christopher Columbus and the Columbian Legacy.* Knopf, New York.

Seeland, K. (1997) *Nature is Culture: Indigenous Knowledge and Socio-Cultural Aspects of Trees and Forests in Non-European Cultures.* Intermediate Technology Publications, London.

Shands, W.E. (1993) National forests and the human legacy: some history. In: Foley, L.H. (ed.) *Silviculture: From the Cradle of Forestry to Ecosystem Management: Proceedings of the National Silviculture Workshop: Henderson, NC.* USDA Forest Service, Asheville, North Carolina, pp. 3–11.

Strathern, M. (1980) No nature, no culture: the Hagen case. In: MacCormack, C. and Strathern, M. (eds) *Nature, Culture and Gender.* Cambridge University Press, Cambridge.

Tobey, R.C. (1971) *The American Ideology of National Science, 1919–1930.* University of Pittsburgh Press, Pittsburgh.

Uhl, C., Nepstad, D., Buschbacher, R., Clark, K., Kauffman, B. and Subler, S. (1990) Studies in ecosystem response to natural and anthropogenic disturbances provide guidelines for designing sustainable land-use systems in Amazonia. In: Andrews, A.B. (ed.) *Alternatives to Deforestation: Steps toward Sustainable Use of the Amazon Rain Forest.* Columbia University Press, New York, pp. 24–42.

Williams, M. (1989) *Americans and Their Forests: a Historical Geography.* Cambridge University Press, New York.

Williams, M. (1999) The end of modern history. *Geographical Review* 89, 275–300.

Forest Management from Positivism to the Culture of Complexity

4

O. Ciancio and S. Nocentini

Dipartimento di Scienze e Technologie Ambientale Forestali, Università di Firenze, Via San Bonaventura 13, 50145 Florence, Italy

The practice of forestry has long been characterized by a linear paradigm. Forest cultivation and management have been centred on the standing volume–regeneration relationship, considered, respectively, as a source of income and as a basis for the continuity of production. This approach is based on the mechanistic concept of the forest viewed as a machine that, if properly arranged and regulated, is capable of providing direct and indirect benefits. The scientific paradigm is Cartesian or Newtonian, and is based on the objectivity of science. The enlightenment culture started the extraordinary development of modern science and technology. The aim was that of emancipating man through his domination of nature. This conceptual apparatus, that did indeed give birth to forestry as a science and has been sublimated by the German school, is today showing its limits. The forest is no longer seen just as trees but instead as a complex biological system with intrinsic values. The implications of the shift from a mechanistic and positivist approach to a systemic and non-linear one are examined in relation to the history of forest management.

1 Introduction

'If there is a field that has remained closed to modern science's positivism, this is truly that of forest management. It is not that attempts have not been made to subject management to strict rules, precise procedures, and if that is where positivism lies, then it is abundant and overabundant.' Thus wrote Henry

(eds M. Agnoletti and S. Anderson)

Biolley in 1920 when he argued in favour of the principles of Adolphe Gurnaud's 'control method'.

Positivism in forestry starts from afar. This chapter examines its influence on the evolution of forest thought and underlines how positivism still strongly influences forestry notwithstanding the development of the culture of complexity and of a new way of looking at the forest.

2 The birth of forestry science: the forest from reserve to resource

In the *Encyclopédie* by Diderot and d'Alembert (1751–1772), the term forest is defined by Le Roy. It is interesting to recall what Harrison (1992) writes about this definition. According to Le Roy the forest is reduced to 'vegetation' or greenery. The forest is nothing more than an agglomeration of trees. Le Roy never speaks about animals. This is because the forest was already conceived of in terms of wood. Wood, in turn, was conceived of in terms of its useful value. Diderot and d'Alembert's modern humanism gave rise to the concept of the forest as a resource, a concept that is the background for the birth and development of forest science.

In line with what Heilbron (1990) calls 'the quantifying spirit of the eighteenth century' forest management at the end of the 1700s turned to mathematics and geometry to organize production. We find the most significant example of this quantifying spirit in Germany. In the era of the 'enlightenment', the improvement of fiscal administration and resource management was seen as requiring a science of state finances, which became known as the 'cameral science' (*Kameralwissenschaft*) in Germany (Tribe, 1980). Forest management was one aspect of state administration that was scrutinized by this new 'science' in order to fit, as Bechstein wrote in 1797, 'scattered pieces of knowledge . . . into systems' and to transform 'all sorts of activities left to habit . . . into a science'. The result was the quantification and rationalization of nature and the regulation of economic practice.

Detailed inventories of the number and species of trees in the forests were drawn up. Plans for cutting and replanting were developed in order to maximize yield while maintaining a constant level of productivity. Methods were studied for estimating the mass of wood in the forest and for calculating the amount of wood used. Geometric formulae were used to represent the shape of the trees, and everyone recognized the utility of a *Normalbaum*, or standard tree, to which the natural specimens could be likened (Lowood, 1990).

The birth of the forestry schools marked a turning point in the way of cultivating and managing the forest. The path was blazed by Johann Heinrich Cotta and Georg Ludwig Hartig who are generally recognized as the founders of forest science – *Fortswissenschaft*. Cotta organized the school of Tharandt in Saxony (1816) and Hartig the school of Hunden in Hesse (1789). Reducing the forest to utility, in a manner similar to intensive farming models, was the main guideline of the German school. 'Scientific forest management' was imported to France from Germany in the 1820s when Bernard Lorentz – a friend of Hartig – founded the forestry school in Nancy.

Even though it may seem paradoxical, it is a fact that the birth of forestry schools and the spread of forest science caused the disappearance of the last remnants of the primeval forest. In Germany, under the impetus of ideas propagated by the schools, natural forests were replaced by geometrically arranged monospecific plantations. The objective of forest management was the pursuit of 'normality'. The forest was built and ordered to supply a high yield of wood. Extensive silviculture, based on the felling of single trees over vast areas, was replaced by intensive silviculture, based on clear felling or on uniform shelterwood felling. This form of silviculture, that some have defined as financial and others as rational, led first to the abandonment and then the disappearance of traditional, empirical silviculture that had been tested through the years and was linked to the family economies of the rural populations.

3 The return to nature

Towards the end of the last century, following on the heels of some blatant failures, opposition to this German form of scientific, financial silviculture started to grow. Karl Gayer with the famous call 'back to nature' (*Zurück der Natur*) and Adolphe Parade with the famous aphorism '*Imiter la nature, hâter son oeuvre*' (imitate nature, hasten its works) are the forerunners of naturalistic silviculture.

Adolphe Gurnaud's presentation (1884, 1886, 1890) of *la méthode du contrôle* (the method of control) was a true revolution. Gurnaud refused to impose a predetermined order to the forest: it was an order that supported the silvicultural concept of the artificial forest. But, even though there was a change in cultivation methods and an attempt to harmonize natural forces, the object still remained that of achieving maximum wood yield. Biolley thus wrote: '"Produire": ce mot renferme tout le devoir et devrait résumer tout le vouloir du sylviculteur' ['To produce': this word encompasses all the duties and should sum up all the desires of the forester].

For a long time forest management remained anchored to the bases of financial and rational silviculture. It is interesting to remember that in 1833 in Baden-Württemberg a law was enacted that prohibited *Femelshlag* and *Plenterwald* (group shelterwood and group selection felling). This law, writes Wolf von Hockenjos (1993, 1995), was considered progressive and liberal and remained in force for 143 years, that is up to 1976. But only in 1992, after 159 years, did the Ministry for Agriculture and Forestry of Baden-Württenberg invite foresters to apply *Femel* and *Plenter* in forest regulation plans.

In 1936 Aldo Leopold, after a visit to Germany and Czechoslovakia, reported that, following spruce cultivation, 'mixed forests and natural regeneration had become relics of an archaic past'. But he also sensed that there was a change in the air. 'Germans are now perceiving that [wood] increment bought at the expense of soil, of the beauty of the landscape and of wildlife is both poor economics and poor public policy' (Leopold 1936a,b).

In 1988, after more than 50 years, Wolfe and Berg re-examined the question. They reported that, contrary to Leopold's predictions, spruce coverage has increased progressively during the last century and now covers 40% of the total land area. Scots pine covers 20%. The prevailing silvicultural system is still even-aged monocultures with clearcut and artificial regeneration. Rotation periods for spruce stands range from 80 to 100 years.

According to Holdgate (1993) the model of 'scientific forestry' was translated virtually unchanged and unchallenged through the curricula of 20th-century forestry schools and colleges. He suggests that the concept of the forest as a timber resource had been exported from Europe to many parts of the world during the colonial era, and left assumptions that persist today, very much to the detriment of the world's forested countries and of those who live in them. According to Twight (1992) 'sustained-yield forestry' still inspires laws and professional values that underlie forestry.

4 Organicism and the *Dauerwald*

The debate between the reductionist view and the organicistic or holistic view is a constant that repeats from one generation to another. The philosophical concept of holism is historically attributed to South African General and Prime Minister Jan Christian Smuts who proposed his views in 1926. A few years earlier in 1916, Frederic E. Clements, in the United States, had defined as superorganism those units of nature that he considered organic, that is with a natural cycle similar to that of a human being. The sequences of vegetation were proof that the biotic community behaved like a complex organism.

In forestry this position was espoused most vigorously by Alfred Moeller with the *Dauerwald* theory. Moeller's thought is summed up in his statement 'I believe I have found the right word. The forest is truly a living organism.' The *Dauerwald* was surely a revolutionary theory. And that is why it was either passionately defended or completely rejected (Dengler, 1930).

In ecology the concept of superorganism has been supplanted by Arthur Tansley's (1935) ecosystem concept. Ecological thought is certainly holistic, but it is also something more. On one hand it underlines the need to look at the whole instead of the parts, and on the other it considers the whole as a part of larger wholes and the relationships between these and the environment.

5 Towards the culture of complexity

Silviculture and forest management refer both to bioecology and economics. They have been developed along the lines of the physicalist method. But now this method is being questioned by the same physicalists because it is not able to provide a significant impetus for understanding living systems. This scientific approach is none the less difficult to remove because it is part of our culture and has been the basis for technological progress.

In recent years the forest has attained a new cultural dimension. It has acquired the status of a complex biological system, that is a system that is autonomous and has the ability to subordinate structural changes to conserving its organization (Ciancio, 1991; Ciancio and Nocentini, 1995, 1996a,b).

The new scientific paradigm is based on the concept of the intersubjectivity of science. In brief, the description of phenomena is also dependent on the observer. The metaphor of knowledge is that of the network of relations. The knowledge process is based on the systemic outlook. The experimental approach is holistic, and the method of trial and error is followed (Ciancio & Nocentini, 1996b) (Fig. 4.1).

6 Systemic silviculture and sustainable forest management

How does forest management fit into this new scientific paradigm? It must be recalled that, starting from the 1970s, the way of looking at the forest has changed. The awareness that the forest plays a fundamental role in the present and future life on our planet has grown. From the famous Club of Rome report *The Limits to Growth* (Meadows, 1972) and from 'A blueprint for survival' (Goldsmith, 1972) to the first Earth Summit in Stockholm in 1972 and to the Bruntland Report (World Commission on Environment and Development, 1987) – known as *Our Common Future* – the urge towards sustainable development has become stronger and stronger.

The World Conference on the Environment and Development, held in Rio de Janeiro in 1992, emphasized placing restrictions rather than limitations on development, shifting attention towards environmental compatibility. Chapter 11 of the Agenda 21 stated the need for scientifically valid criteria and guidelines for sustainable forest management, conservation and development. The Conference for Forest Protection in Europe, held in Helsinki in 1993, ended with four resolutions that are a formal commitment for European countries.*

---

*The four resolutions are: H1 'General guidelines for sustainable forest management in Europe'; H2 'General guidelines for biodiversity conservation in European forests'; H3 'Forest cooperation with countries with transition economies'; and H4 'Strategies for a long-term adaptation process of European forests to climate changes'.

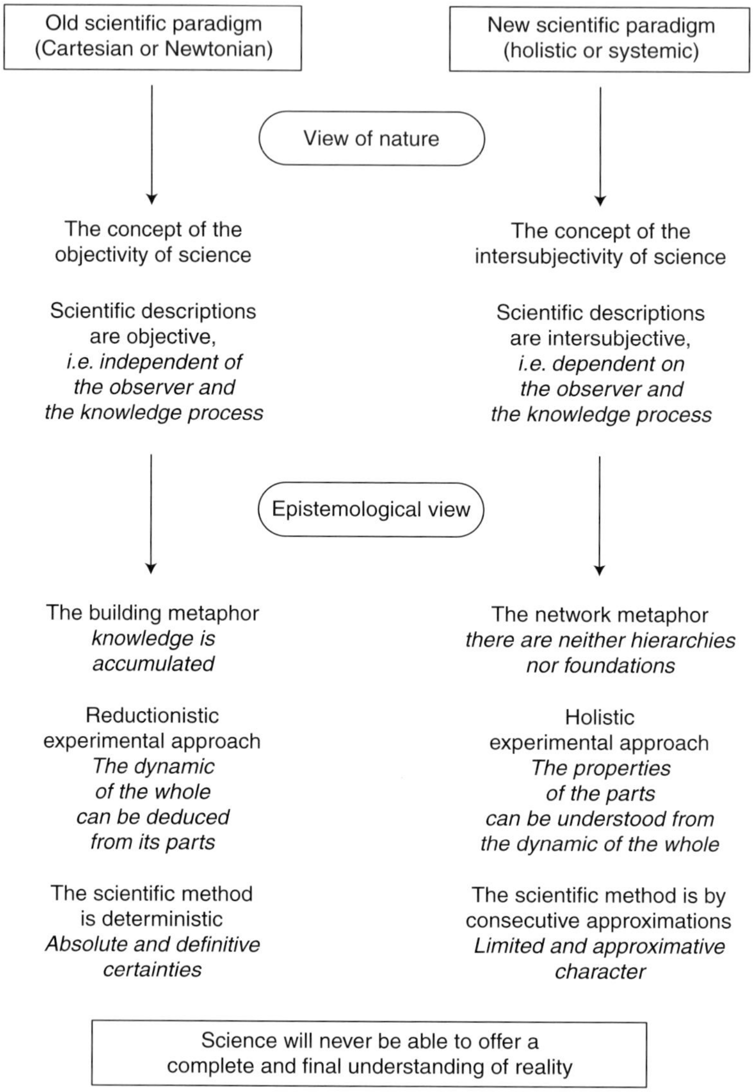

Fig. 4.1. The old scientific paradigm and the new scientific paradigm.

The elaboration of ecological thought has given rise to an important debate on environmental issues and has generated a series of cultural movements that have promoted, among other things, the critical review of what can be termed the 'forestry question' (Ciancio, 1992). A crucial point is tied to the conflict between ecology and economics. There is a growing awareness that natural resource problems can be solved only under the condition that ecology and economics be confronted with ethics. As Gregg (1992) has pointed out, foresters have preferred to look to forest science for the resolution of practical forestry problems. But the current debate is demonstrating the inadequacy of a solely technocratic approach. As long as words such as 'sustainability' and 'biodiversity' are not attached to 'values' they will continue to serve more as evasive indicators than as practical answers to real questions. Biodiversity has cultural and use value because it enhances 'local knowledge' that is guarded by the local communities that live in the forest.

The forest should not be viewed as a completely disposable asset and should not be managed only according to the principles of market economy (Ciancio and Nocentini, 1996a). As Georgescu-Roegen (1976) has said, market alone causes a greater consumption of resources by the first generation, and more rapid consumption than appropriate.

7 The classic forestry system and the autopoietic forestry system

McQuillan (1998) maintains, with the new chaos and complexity theories, we can no longer share the optimism of 19th-century science since the future trajectories of all complex systems are inherently unpredictable. For this reason we must learn to live actively yet prudently, aware of being in a state of risk, uncertainty and ignorance (Faber *et al.*, 1992). Thus, the question is: How can we transform this awareness into concrete actions? With the current state of knowledge, due to the indeterminism mentioned earlier, we cannot reliably predict the outcome of all decisions and the consequences of silvicultural interventions or actions. They are not performed in laboratories and they cannot be simulated on a computer. They must be carried out within the ecosystem in the forest, and they may lead to errors which in turn become environmental, economic and social burdens.

To avoid problems of this type, we must make flexible decisions that can discover any errors and promptly correct them at the lowest possible cost. In brief, we must reconsider the positivist approach that still dominates some academic and research sectors, and adopt the 'trial and error' (adaptive management) approach that is based on a scientific attitude. This is possible if we adopt a management form that considers the forest to be an autopoietic system, that is a system capable of self-organization and of combining functional efficiency with high technical and economic value along with ecological and cultural values (Ciancio and Nocentini, 1995, 1996a; Ciancio, 1998).

Table 4.1 shows a comparison of the structural features of the classic (SF_1) and the autopoietic (SF_2) forestry systems, along with the impact of management, ecological evaluations and the intended primary objective. The classic forestry system (SF_1), anchored to the mechanistic paradigm, is a linear system. It is a system based on the principle that the forecasts of results in

Table 4.1. A broad comparison of the classic forestry system (SF_1) and the autopoietic forestry system (SF_2).

Classic forestry system (SF_1)	Autopoietic forestry system (SF_2)
Linear system, lacking in alternatives	Non-linear system, rich in alternatives
Uniformity and homogeneity of the system	Lack of uniformity and homogeneity of the system
Reduction of diversity and loss of genetic information	Diversity is a source of genetic information, it has cultural and useful values
Management	
Cultivation uniformity demands centralized control in relation to profit and market	Cultivation diversity requires decentralized control and properly exploits local 'knowledge'
Rigidly structured forest in age or diameter classes	Unstructured forest capable of 'self-organization'
Product uniformity: primarily wood (PR_1)	Diversified products: including wood (PR_2)
Ecological evaluation	
Stable and sustainable system with input of energy, labour and capital. Productivity, yield and economic value are independent of the ecosystem	Stable, sustainable and autonomously renewable system. Productivity, yield and economic value are dependent on the ecosystem
Primary objective	
To maximize profits through commercial use of the forest	To increase complexity and conserve biodiversity

terms of wood production are linear. This means that, within certain limits, the yield increases proportionately to the energy used. Such a forestry system must follow reference standards that may be identical for different forests or differentiated within given limits; thus management alternatives are limited. A linear system means a silvicultural orientation that tends towards a uniform and homogeneous forest and therefore a reduction in biodiversity. A forest managed according to such criteria is almost always monospecific or it consists of a main species and one or, at most, two secondary species. Its structure is rigidly even-aged or uneven-aged and it is arranged chronologically or according to diameter.

Forestry planning is anchored to the theory of economic development which, in turn, is firmly based on exponential growth models. Management plans are evaluated on the basis of their ability to sustain a high rate of wood production. The rotations are relatively short with respect to the longevity of the species. This type of management presupposes centralized control and a generic crop uniformity.

The classic forestry system (SF_1) is a stable and sustainable system as far as wood production, landscape and recreational use are concerned. However, believing that the 'wake' or *Kielwassertheorie* effect, related to management focused on production, is capable of satisfying the other needs is an arbitrary, scientifically irrelevant simplification that is unacceptable in so far as ecology and biodiversity are concerned. Management – which is sustainable within a given space of time – systematically tends to maximize the soil rent. This makes it unlikely, if not impossible, that the ecosystem will function. It emphasizes wood production to the detriment of other factors indispensable for the system's efficiency. In brief, productivity, yield and economic value are independent of the ecosystem, while sustainability is dependent upon the input of energy, work and capital. This means that production is tied to a high level of external input. And this, since it markedly affects the ecosystem's equilibrium, leads to high vulnerability and ecological instability, significant erosion of biodiversity, lack of alternatives and a low option value.

The autopoietic forestry system (SF_2), on the contrary, is a 'non-linear' system. It is rich in biodiversity. It is capable of providing alternatives; since it does not follow reference standards, it varies over the short term and adapts to different situations. A system of this type involves an orientation that aims at conserving or increasing biodiversity, hence less uniformity and less homogeneity: the structural complexity of the forest. Biodiversity has cultural value and use value because it makes it possible to properly exploit 'local knowledge', the property of communities that live with the forest, and to obtain products that differ from valley to valley. Wood is an important product, but it is not the sole objective of management.

Forestry planning is anchored to systemic management. The silvicultural unit is the stand. Actions are based on the needs of the stand and are undertaken cautiously, continuously and in a capillary manner. Their objective is active participation in the evolutionary processes of the ecosystem. Verification of feedback and reactions to the interventions is an essential aspect of management. During planning, the coordinates for the operational decisions do not take common parameters such as rotation or diameter class into consideration. Systemic management means decentralized control and crop diversification.

The autopoietic system (SF_2) is capable of satisfying society's needs and of properly exploiting the wealth of experience of the local communities. This type of management aims at achieving functional efficiency of the ecosystem and orienting silvisystems towards environmental integrity. Systemic management leads to sustainability because it enhances the potential of distributing the forest's multiple services and products. Productivity, yield and economic value are dependent upon the ecosystem. Sustainability is independent of the input of energy, labour and capital. This means that production is tied to a low level of external input. And since this does not significantly impact upon the ecosystem's equilibrium, it leads to high ecological stability, the ability to conserve or augment biodiversity, a wealth of alternatives and a high option value.

8 Conclusion

It must not be forgotten that forestry as a science was born and developed in a period when the dominating culture was founded on deterministic predictability. Pierre Simon de Laplace (1814) wrote:

> An intellect that, in a given moment, could know all the forces that animate nature and the situation of each being that composes it, and is sufficiently vast to analyse all these data, such an intellect could embrace in the same formula the movements of the greatest bodies of the universe and that of the lightest atom; nothing would be uncertain for him and the future, like the past, would be present in his eyes.

Today, after the theory of relativity, the quantum theory, the principle of indetermination and the chaos theory, the dream of Laplace seems, at the least, inconsistently optimistic. Nevertheless, faith in the ability of man to dominate nature, and to predict and control events remains the main object of modern science.

The new scientific paradigm does not find the necessary consensus in the academic and technical world. This is due to the fact that there still is a positivist culture that considers science as separate from ethics. But, in Marcuse's (1966) words, 'a science separated from ethics results in a science devoid of art and aesthetics'. In biology, scientific explanations are sought through analysis that is an expression of continuum and not through the examination of discontinuum that is typical of all systems. One of the main points of the positivist culture was superseded when it was understood that complexity cannot be interpreted and explained with this type of language.

We will end with what Max Plank used to say, recalling Galileo's words: the triumph of a new scientific truth is not due to the fact that it is able to convince those who oppose it, but only to the fact that at last its opponents die and a new generation grows up that is familiar with it. That is to say that, as always, today's frontiers are the limits of tomorrow.

References

Bechstein, J.M. (1797) Anzeige von der Herzoglich-Sächsisch-Gothaischen und Altenburgischen Societät der Forst- und Jagdkunde zu Waltershausen nebst den vorläufigen Statuten derselben. *Diana* 1, 424–429.

Biolley, H. (1920) *L'aménagement des Forets par la Méthode Experimentale et Spécialement la Méthode du Controle* [*Forest management according to the experimental method and especially the control method*]. Attinger Frères, Neuchatel.

Ciancio, O. (1991) La gestione dei querceti di Macchia Grande di Manziana: la teoria del sistema modulare [Management of the oak stands of the Macchia Grande di Manziana: the theory of the modular system]. *Cellulosa e Carta* 42, 31–34.

Ciancio, O. (1992) La questione forestale italiana: l'orizzonte possibile [The Italian forestry question: the possible horizon]. *L'Italia Forestale e Montana* 47, 321–339.

Ciancio, O. (1998) Gestione forestale e sviluppo sostenibile [Forest management and sustainable development]. In: *Proceedings of the Second National Congress of Silviculture, Per il miglioramento e la conservazione dei boschi.* Venice, 24–27 June, 1998. Consulta Nazionale per le foreste e il ligno, Direzione generale per le risorse forestali, montane e idriche, Roma, Vol. 3, pp. 131–187.

Ciancio, O. and Nocentini, S. (1995) Nuovi orientamenti in selvicoltura [New trends in silviculture]. In: Georgofili, I. (ed) *Atti dell'Accademia dei Georgofili*, Vol. XLII, Florence, pp. 11–153.

Ciancio, O. and Nocentini, S. (1996a) The forest and man: the evolution of forestry thought from modern humanism to the culture of complexity, systemic silviculture and management on natural bases. In: Ciancio, O. (ed.) *The Forest and Man.* Accademia Italiana di Scienze Forestali, Florence, pp. 21–114.

Ciancio, O. and Nocentini, S. (1996b) The scientific paradigm, 'good silviculture' and the wisdom of the forester. In: Ciancio, O. (ed.) *The Forest and Man.* Accademia Italiana di Scienze Forestali, Florence, pp. 259–270.

Clements, F.E. (1916) *Plant Succession: Analysis of the Development of Vegetation.* Carnegie Institute, Washington, DC.

Dengler, A. (1930) *Waldbau auf ökologischer Grundelage* [*Silviculture on ecological bases*]. Julins Springer, Berlin.

Diderot, D. and d'Alembert, J.B. (1751–1772) *Encyclopédie ou Dictionnaire raisonné des Sciences, des Arts et des Métiers, par une Société de gens de lettres.* Briasson, David, Le Breton, Durand, Paris.

Faber, M.R., Manstetten, R. and Proops, J.R.L. (1992) Humankind and the environment: an anatomy of surprise and ignorance. *Environmental Values* 1, 217–242.

Gayer, K. (1880) *Der Waldbau* [*Silviculture*]. Wiegandt, Hempel, Parey, Berlin.

Georgescu-Roegen, N. (1976) *Energy and Economic Myths.* Pergamon Press, New York.

Goldsmith, E. (1972) A blueprint for survival. *The Ecologist* 2, 1–43.

Gregg, N.T. (1992) Sustainability and politics: the cultural connection. *Journal of Forestry* 90, 17–21.

Gurnaud, A. (1884) *La Sylviculture Française* [*French sylviculture*]. Jacquin, Besançon.

Gurnaud, A. (1886) *La Sylviculture Française et la Méthode du Contrôle* [*French sylviculture and the control method*]. Jacquin, Besançon.

Gurnaud, A. (1890) La Méthode du Controle et la Tradition Forestière [*The control method and forestry tradition*]. *Revue des Eaux et Forets.*

Harrison, P.R. (1992) *Forests: the Shadow of Civilization.* University of Chicago Press, Illinois.

Heilbron, J.L. (1990) Introductory essay. In: Frangsmyr, T., Heilbron, J.L. and Rider, R.E. (eds) *The Quantifying Spirit of the Eighteenth Century.* University of California Press, Berkeley, pp. 1–23.

Holdgate, M. (1993) Sustainability in the forest. Keynote address to the conference. *Commonwealth Forestry Review* 72(4), 217–225.

Hockenjos, von W. (1993) Die Wiederentdeckung des Femelwaldes [The return to the Femelwaldes]. Auf forstgeschichtlicher Spurensuche im Bücherschrank eines badischen Forstamtes. *Allgemeine Forst und Jagdzeitung* 164(12), 213–218.

Hockenjos, von W. (1995) Forstideologisches aus Baden [Ideological background of Badens forestry]. *Allgemeine Forst und Jagdzeitung* 166(2/3), 34–38.

Laplace, de P.S. (1814) *Essai Philosophique sur les Probabilités*, 2nd edn. Paris. (Anastatic reprint, Culture et Societé Bruxelles, 1967.)

Leopold, A. (1936a) Deer and Dauerwald in Germany. I. History. *Journal of Forestry* 34(4), 366–375.

Leopold, A. (1936b) Deer and Dauerwald in Germany. II. Ecology and policy. *Journal of Forestry* 34(5), 460–466.

Lowood, H.L. (1990) The calculating forester: quantification, cameral science and the emergence of scientific forestry management in Germany. In: Frangsmyr, T., Heilbron, J.L. and Rider, R.E. (eds) *The Quantifying Spirit of the Eighteenth Century.* University of California Press, Berkeley, pp. 315–342.

McQuillan, A.G. (1998) Defending the ethics of ecological restoration. *Journal of Forestry* 96(1), 27–31.

Marcuse, H. (1966) *One Dimensional Man.* Beacon Press, Boston, Massachusetts.

Meadows, D.H. (ed.) (1972) *The Limits to Growth.* Universe Books, New York.

Moeller, A. (1922) *Der Dauerwaldgedanke. Sein Sinn und seine Bedeutung* [*The Dauerwald. Its meaning and its importance*]. Springer Verlag, Heidelberg.

Smuts, J.C. (1926) *Holism and Evolution.* MacMillan, New York.

Tansley, A.G. (1935) The use and abuse of vegetational concepts and terms. *Ecology* 16, 284–307.

Tribe, K. (1980) Cameralism and the science of government. *Journal of Modern History* 9, 172–196.

Twight, B.W. (1992) Sustained yield forestry: a source of conflict in environmental protection? In: *Proceedings of IUFRO Centennial. Division 6 and 4.* IUFRO Secretariat, Vienna. pp. 58–68.

Wolfe, M.L. and Berg, F.C. (1988) Deer and forestry in Germany, half a century after Aldo Leopold. *Journal of Forestry* 86, 25–31.

World Commission on Environment and Development (1987) *Our Common Future.* Oxford University Press, Oxford.

Economic Areas and Forest Nature: the Search for Forest Images and the Understanding of Nature in the Past

5

B. Selter

Forstliche Dokumentationsstelle der Landesforstverwaltung, Nordrhein-Westfalen, Herbreme 2, D-59821 Arnsberg, Germany

There have always been different opinions of what a forest should be and what appearance it should have. The present forest environment is the result of historical development. When we are talking about woods and their resources, about man exploiting forests over centuries, it is also important to analyse environmental behaviour and ecological awareness in history. In many ways the image of nature in previous ages was a quite different one to nowadays. Woodlands served other functions for peasants than for foresters or today's vacationers. Humans defined nature by means of their activities in nature. Woodlands were economic areas. Because of its sources, forest history puts us in a position to obtain a whole lot of information about forms of former ecological crises (deforestation, wood shortage) as well as about the complex links between people and nature. The knowledge of historical presuppositions can support contemporary management of forests. We have to formulate questions on the stability of former forest exploitation systems and the imaginable existence of anthropological constants in forest usage. This complex of questions should be part of a broader conception of environmental history that is also relevant for forest history.

1 Introduction

There have always been different opinions of what a forest should be and what appearance it should have. Present forest environment is the result of historical development. When we are talking about woods and their resources, about man exploiting forests over centuries, it is also important to analyse environmental behaviour and ecological awareness in history. Many questions are still unanswered.

Because of its sources, forest history puts us in a position to obtain a whole lot of information about former ecological crises as well as about the complex links between people and nature. In the following I will raise some of the above problems and give a short insight into previous human relations with the limited forest resources in the south Westphalian hilly region. Far into the 19th century forests were subordinated to agricultural and industrial interests in this region. After the 1860s the technological changes of the Rhenish-Westphalian iron industry defined new forest functions. The demands for timber changed and the transition from production of firewood to that of commercial timber began. This involved a drastic change of wood species.

2 Views of woodlands and forests

Most of us use terms like 'nature' or 'environment' vaguely and synonymously. Moreover, there has been no acceptable definition of 'environment' in environmental history up to this day. Its meaning is always overlaid by human attitudes and activities which strive to live in harmony with an ideal nature.

When we see environment only as a biosphere separated from human beings, we do not leave any room for a nature that meets us almost as a social constituent. Many historians equate 'nature' with 'environment'. From an anthropocentric point of view, environment is seen as nature for humans. We cannot take a precise meaning of 'nature' and 'environment' for granted, considering the fact that the meaning of both terms has been subjected to many transformations in history. 'Humans change nature by cultivating activities and with it its meaning,' as Sieglerschmidt (1994) said.

Along with an endangered nature we also see our own existence endangered. That is why threatened nature deserves our interest. In the past, things were not really different from today, but people stressed other issues. Thus human perception of nature and forests could vary, depending on different social groups. It extended from the so-called (in German) *Nährwald* over the efficiently cultivated 'timber fields' to forest idylls, protected for aesthetic reasons.

When, in the 19th century, thinkers and cultural critics adopted the romantic–idealistic transfiguration of nature, in many regions of Germany the forest images decisively changed. In Westphalia, too, the first artificial forests were covering wide areas. Against that the educated classes put forward the ideal of unspoiled nature and the improvement of the landscape and produced the idea of care and protection of natural monuments (*Naturdenkmalpflege*). Associated with antimodernistic criticism of civilization they created their own forest nature as a kind of counterpart and recreation area.

Misjudging the real situation, they thought they would find this kind of nature in the forest's grazings lands and the selection forests that had already been changed by human activities. Demand and reality of this early form of landscape protection diverged widely. Nevertheless, the technical exploitation of forests and the extensive afforestation promoted the idea of forest protection. At the end of the century the nature conservation movement was founded in Germany. People practised landscape protection in order to conserve scenic characteristics and to preserve native habitats.

Forestry science of that time hardly accepted this social–romantic view of woodlands. It also did not find its way into the thought and behaviour of the rural population. For both groups woodlands were economic areas, part of the natural resources available for human beings. The sustained-yield system was the theoretical argument for the increasing reduction of forest functions to the economically usable quantities of wood. The so-called 'normal forest' was intended to regulate 'disorderly nature' with the help of science. In Westphalia this was the prelude to a long-term remodelling of the landscape originally covered with hardwood by laying out monotonous pine and spruce cultures (Fig. 5.1).

3 Points of contact between environmental history and forest history: questions and research assignments

- What kind of forest image finally induced humans to act?
- On what scale did these patterns of behaviour affect social groups using the forests?
- Can we possibly find any ecological awareness 100 or 200 years ago?

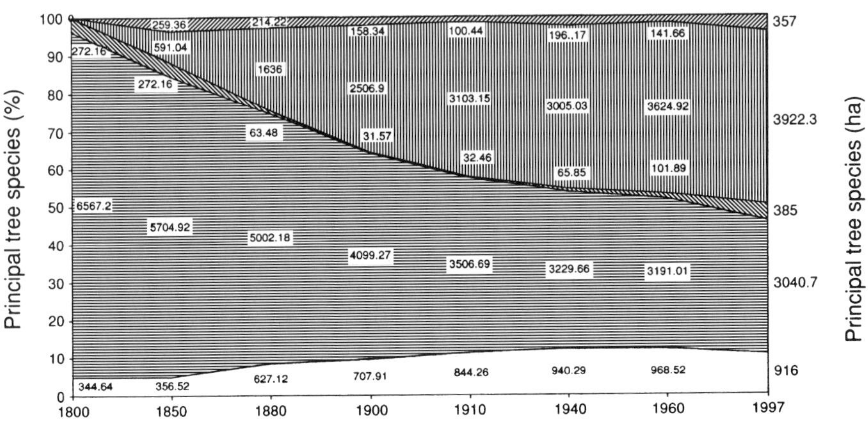

Fig. 5.1. Change of wood species in the Arnsberg Forest from 1800 to 1997. ☐, Oak; ▤, beech/other tree species, low rotation; ▧, older/other tree species, high rotation; ▥, spruce/Douglas fir/fir; ▨, pine, larch. Source: Records Staatliches Forstamt Arnsberg.

- Did humans perceive changes of environment, and in what ways did they react to them?
- Can we regard earlier forms of environmental pollution and deforestation as a decisive part of the social development to a throw-away mentality?
- In what way did people live with limited resources?
- Can the integrated whole of agriculture, preindustrial industry and utilization of forests form a stable system which does not endanger forest stands?

Along with the exploration of so-called endogenous factors – that is the natural and material environment – a broader conception of environmental history is necessary. A kind of cultural history of forest exploitation is conceivable. This would include both intentional and non-intentional human influences on woodlands (that is exogenous factors) – including the effects on environment and the affected social systems.

Let me outline a short history of forest utilization in southern Westphalia divided up into different periods. The first phase, until the beginning of the 19th century, was formed by the preindustrial agrarian society. The close interweaving between traditional agriculture and largely unregulated forestry was characteristic. Both ranged – as Sieferle (1993) argues – 'on principle between the stationary limits of shortage of the natural solar energy system'. In some regions a shortage of wood occurred, but no perception of damage to the environment existed. Since the causes of interrelations between the utilization of forests and possible deforestation were not known, humans mostly drew conclusions in connection with the endangering of their own existence. The rulers/sovereigns exploited the anticipation of wood shortage for their specific aims. The dealing with nature was more economical, and this period showed phases of regeneration.

The second phase came to an end in the late 19th century. It was the beginning of a restraining of agricultural forest uses step-by-step. External input rates (e.g. chemical fertilizers) were not used. But we can detect a noticeable intensification of material cycles and flows (transfer of nutrients from woodlands to the fields by wood pasture and forest litter utilization). In industry, the conversion to fossil energy sources started and the rates of external input increased. The result was a regional, extensive lack of timber and clear devastation of many wooded areas. The present or impending scarcity of wood put man under an increasing pressure of adaptation. Economic incentives (*Anreize*) were caused by the expanding commercial timber market. Both encouraged the willingness to change existing structures (reafforestation, changes of agrarian structures).

Only at the beginning of our century, within the third phase, was a timber economy initiated in all forest estates. The process of the separation of agriculture from forestry was largely finished. Both agriculture and industry were working with external inputs. Timber shortage declined because the increasing needs for lumber were covered by imports. Nowadays a new type of forest damage occurs in the advanced stage in this process.

4 The exploration of mental–intellectual processes

It is very difficult to track down a concept of environment or a kind of preindustrial ecological awareness. The records mention nothing about it: they rarely reflect the opinion of the rural population anyway. The aesthetic experience of nature in particular makes visible to us the mental distance between the authors of forest administration files and the people living with nature. In the country the alienation from nature did not exist in the same way as in the towns. Humans defined nature by means of their activities in it. Woodlands were economic areas, fulfilling different functions for peasants than for foresters. After 1850 the extensive afforestation with spruce and pines resulted in a lasting disturbance of the traditional mechanisms of regulation between agriculture and forest economy in southern Westphalia. Opposition to this was the order of the day.

People made no protest against afforestation with coniferous wood for fear of possible ecologically negative effects. They were worried about the subsidiary uses of forests. Many of these uses could not be practised in coniferous forests. The awareness of possibly existing damage to the environment here shows us a selective structure: the satisfaction of the most elementary needs allocates environment a lower place in the scale of preference. A threat of wood scarcity was decisive, not the future increase of forest areas.

But was this 'unecological' behaviour? That is difficult to find out. The experience of many centuries determined the use of fields and woodlands, not an unhesitating overexploitation. There were no alternatives to fertilizing with wood ashes or hacking heather. For a long time a meagre base of feed had forced the farmers to use woodlands as a substitute for pasture. In South Westphalia agricultural structures could not adapt to the pace of the new forest economy. Nature set the limits for human work routines. Experience in using forests and wood, gained over generations, was used for the economic cover. Considerations for the growing stock reserves and forest conditions arose from these principles, and less from insights into ecological interrelations and least of all from aesthetic emotions.

The forest images of the past were associated with ideas unusual for present-day standards. The ideal forest for a forester aimed at the new timber economy was different from the one peasants used in the 19th century. Their model of woodlands was formed by coppice systems and pasture grounds. In the discussions on responsibility for the historically provable forest damage through overexploitation, environmental historians have to free themselves both from thought patterns of a selfish peasant intentionally wasting forests,

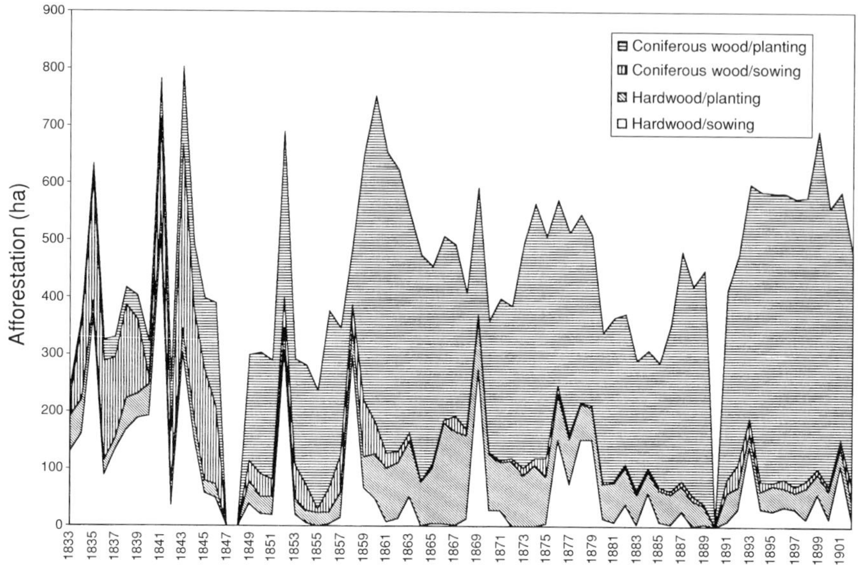

Fig. 5.2. Afforestation in communal forests of the administration district/County of Arnsberg from 1833 to 1902, in hectares. Source: Amtsblätter der Königlich Preußischen Regierung zu Arnsberg. Arnsberg 1834–1903.

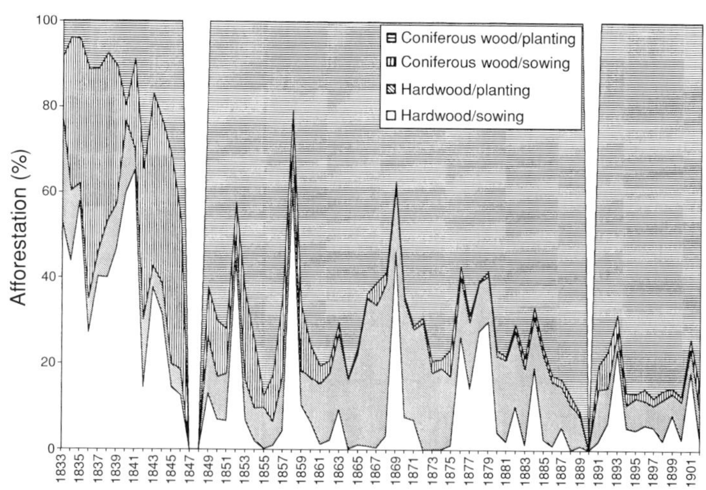

Fig. 5.3. Afforestation in communal forests of the administration district/County of Arnsberg from 1833 to 1902, in percentages. Source: as above, Fig. 5.2.

and from the idea of the forester doing nothing but banishing traditional usufructuary rights from forests.

A further characteristic of forest history is the exploration of unintended long-term effects of human activities. The long circulation periods in silviculture postpone the results of personal actions far into the future, so that only later generations can react to possible damage. Changes of environment are hardly perceived, because such processes happen very slowly. On the other hand, processes of deforestation could still show an exponential development.

It is true that opinions on forest conditions and forestry were not based on an ecological awareness in a present-day understanding. But we cannot deduce from this that there was no kind of conservation in preindustrial times. It is hard to say whether humans correctly judged the capability of forest reproduction. The social dimension of perception was particularly decisive; technical knowledge, however, was limited. The awareness of damage to the environment entailed by the overexploitation of forests only existed in so far as these things were seen as a problem of the limitation of concrete needs. People did not imagine a separate right of nature.

We cannot directly prove an ecofriendly behaviour of historical actors derived from that. But terms like 'devastation' or 'destruction' of forests show us a certain understanding of ecological crises. By the way, not only the forest administration used these terms as an argument, they were also used by the people to achieve their rights of use.

The appearance of the term 'wood scarcity' does not necessarily indicate a real physical situation of shortage. It could refer to a scarcity which came about because of missing funds and means of transportation. 'Conventional' forest history concentrated more on the economically useful high forest and correspondingly welcomed the reforms of forestry in the 19th century. Recent studies are more differentiated. Social and economic history critically interprets forest laws and points out the selfish activities of the big landowners and industrial forest users. Nevertheless, the repeatedly proved identification of forest political and forest economic measures with the interests of special social groups should not lead us to devalue the silvicultural contents of sources in general. The mere exploitation of complaints about wood shortage by no means refutes the real existence of this shortage.

- To what extent was this shortage anticipated at all?
- Was there a mentality of forest utilization in the course of time, approximately at the beginning of the age of mineral coal?
- Did there exist ethics of saving resources up to this time?

In research there is a controversial discussion. While, for example, Radkau (1991) is inclined to think that the throw-away mentality does not constitute an 'anthropological constant' and that overexploitation is not a historically unavoidable necessity, other authors like R.P. Sieferle take a different view. They support the thesis that already in the past 'consumers used goods in the way car drivers do it today' (Sieferle, 1988). 'Stability is a possible condition of ecosystems, however not a standard; that means, there is no ideal or normative ecological balance' (Brüggemeier, 1992). When the actions of man come into play instabilities increasingly occur.

We have to formulate questions on the stability of former forest exploitation systems and the imaginable existence of anthropological constants in forest usage. This complex of questions should be part of a broader conception of environmental history that is also relevant for forest history.

References

Brüggemeier, F.-J. (1992) Umweltgeschichte – warum, wozu und wie? Überlegungen zum Stellenwert einer neuen Disziplin. In: Isenberg, W. *et al.* (eds) *Historische Umweltforschung. Wissenschaftliche Neuorientierung – aktuelle Fragestellungen.* Bergisch Gladbach, pp. 9–26.

Radkau, J. (1991) Unausdiskutiertes in der Umweltgeschichte. In: Hettling, M. *et al.* (eds) *Was ist Gesellschaftsgeschichte? Positionen, Themen, Analysen.* München, pp. 44–57.

Sieferle, R.P. (1988) Perspektiven einer historischen Umweltforschung. In: Sieferle, R.P. (ed.) *Fortschritte der Naturzerstörung.* Frankfurt/M., pp. 307–368.

Sieferle, R.P. (1993) Die Grenzen der Umweltgeschichte. In: *GAIA* 2(1), 8–21.

Sieglerschmidt, J. (1994) Umweltgeschichte und Öffentlichkeit: Probleme der Vermittlung von Umweltgeschichte. In: Winiwarter, W. *et al.* (eds) *Umweltbewältigung. Die historische Perspektive.* Bielefeld, pp. 170–177.

The Age and Size of Hazel (*Corylus avellana* L.) Stools of Nåtö Island, Åland Islands, SW Finland

6

C.-A. Hæggström

Department of Ecology and Systematics, PO Box 7, University of Helsinki, FIN-00014, Finland

The age and size of 84 hazel shrubs (stools) were studied on the island of Nåtö in the southwesternmost archipelago of Finland. The basal girth of all stools and the age and girth of 169 aerial stems of 56 stools were measured. The girth of the thickest aerial stem studied was 73 cm; its age was only 50 years. About 15 shrubs had stems with a girth of at least 40 cm. The oldest stem determined was 60 years, its girth being 45 cm. A growth index for the aerial stems was determined by dividing the girth of the stem with its age. The basal girth of the 84 stools studied varied between 22 and 861 cm. In total, 18 of the studied stools had a basal girth over 600 cm. The maximum number of aerial stems of one stool was about 170. The age of the individual hazel stools was calculated in two ways: (i) as the potential age of the locality of the shrub as a suitable spot for hazel growth; and (ii) as the age of the shrub (stool) calculated according to the growth index. The calculated age of the shrubs and the age of the localities follow each other rather well up to about 5 m above the mean water level, especially when the land uplift rate of 4.5 mm year^{-1} was used in the calculations. The calculated age (according to the mean growth index of 0.87) of the 84 stools was between 25 and 990 years. Due to the methods used, some of the shrubs became older than the age of their locality, especially when the land uplift rate of 5.0 mm year^{-1} was used. This is due to several sources of error which are discussed.

1 Introduction

> There is a lot of romantic guesswork and little hard information about how long trees live and how old particular individuals are. (Rackham, 1983)

A hazel shrub starts from a seed. As a young shrub it is shaped as a bouquet of a few aerial branches (shoots or stems) from its underground stool. The stool grows slowly at its periphery and begins with age to decay in the centre. The aerial stems are rejuvenated from the peripheral parts of the stool. Thus the hazel shrub becomes more or less ring-like with a hollow centre. Such a ring-like hazel shrub is called by a Swedish vernacular term, *runna* (Palmgren, 1915–1917).

The age of individual aerial stems is easy to determine, either on cut stems or on cores taken by an increment core borer. The age of the whole shrub is, on the other hand, difficult, or even impossible, to determine. However, in land uplift areas with a known land uplift rate, such as the Åland Islands in SW Finland, it is possible to calculate the age. This is because the spot on which a certain hazel grows has a defined period of being suitable for hazel growth. In order to determine the age and size of individual hazel shrubs, studies were made on 84 hazels on Nåtö Island in the Åland Islands, in southwesternmost Finland.

2 Materials and methods

Hazel shrubs were chosen according to the following scheme: (i) shrubs growing as near to the sea shore as possible so that the lowest elevation for hazel growth could be established; (ii) shrubs at different elevations, up to the highest point of the study area; and (iii) large shrubs, if possible the largest, in different parts of the study area.

The location of each hazel above the mean water level (MWL) of the Baltic Sea off Nåtö was determined by a levelling instrument. This could be done with an accuracy of about 1 cm. The lowest altitude at which a hazel was found was 83 cm above the MWL. As this shrub was already of some age, the lowest altitude at which a hazel nut supposedly can germinate was therefore arbitrarily determined to be 70 cm above the MWL. The corrected altitude of each locality was thus obtained by subtracting 70 cm from each measured altitude. This corrected altitude was used in the calculations of the age of each locality, thus resulting in the corrected age of the locality.

The height of the shrub was estimated by eye, using the 3 m levelling ruler as a measure. The basal girth of each shrub was measured by a string and a measuring tape. The numbers of living aerial stems were counted and the girth of the thickest was measured with a measuring tape at 40–50 cm above the ground. The girth is a better measure than the diameter, because numerous stems are oval in transverse section. A core was taken from one to seven stems per shrub at 40–50 cm height with an increment core borer. These cores were taken from 56 shrubs. The age of these cores, 169 in all, was determined with the aid of a stereomicroscope. A growth index for the age calculations was determined by dividing the girth of the stem with its age.

In the calculations, a land uplift rate of both 5.0 mm year^{-1} (Witting, 1943; Lisitzin, 1964) and 4.5 mm year^{-1} (Winterhalter *et al.*, 1981) was used. The rate of land uplift is assumed to have been even during the last 2500 years, the period during which the terrain studied emerged from the sea.

The age of the individual hazel stools was calculated in two ways: (i) as the potential age of the locality of the stool as a suitable spot for hazel growth; and (ii) as the age of the stool calculated according to the growth index. The nomenclature of the vascular plants is according to the *Field Flora of Finland* (Hämet-Ahti *et al.*, 1998).

3 Study area

The study area belongs to the Nåtö–Jungfruskär Nature Reserve and is located on the northeastern part of Nåtö Island (Fig. 6.1). The Swedish name Nåtö means the 'Nut Island'. Accordingly, dense stands of hazel (*Corylus avellana*) cover large areas of the island. Nåtö is further renowned for its luxuriant vegetation rich in vascular plant species (Palmgren, 1915–1917; Hæggström, 1983, 1988, 1996).

This study was carried out mainly in the sampling areas 1, 3, 4, 5 and 8 (Hæggström, 1983). A description of the areas is given by Hæggström (1983). The hazel stands are former coppices. The stems were cut frequently, particularly for barrel hoops for salt herring barrels (Palmgren, 1915–1917; Henriksson, 1966); Nåtö was known for its barrel hoop poles of excellent quality. Fishermen from the east Alandian communes came to Nåtö during the latter part of the 19th and early 20th centuries to acquire this quality product for their needs.

The distance between the hazel shrubs varies from 1 to 6 m, being mostly about 2–4 m (Figs 6.2 and 6.3). The height of the old hazel shrubs is mostly 6–8 m although some tall shrubs are about 9 or 10 m high. The individual stems are often longer than the height of the shrub, because they are bent outwards (Palmgren, 1915–1917). The shrubs have a varied number of stems growing from the underground stool.

The vegetation of the hazel stands is luxuriant and rich in species (Hæggström, 1983). The hazel stands are mixed with trees and other shrub species. The most abundant species are *Fraxinus excelsior* (mainly old pollards) and *Betula pubescens*, while *Acer platanoides*, *Alnus glutinosa*, *Betula pendula* (old pollards), *Malus sylvestris*, *Sorbus aucuparia* and *Ulmus glabra* occur less abundantly. *Juniperus communis*, *Ribes alpinum* and *Rosa dumalis* are the most abundant shrub species.

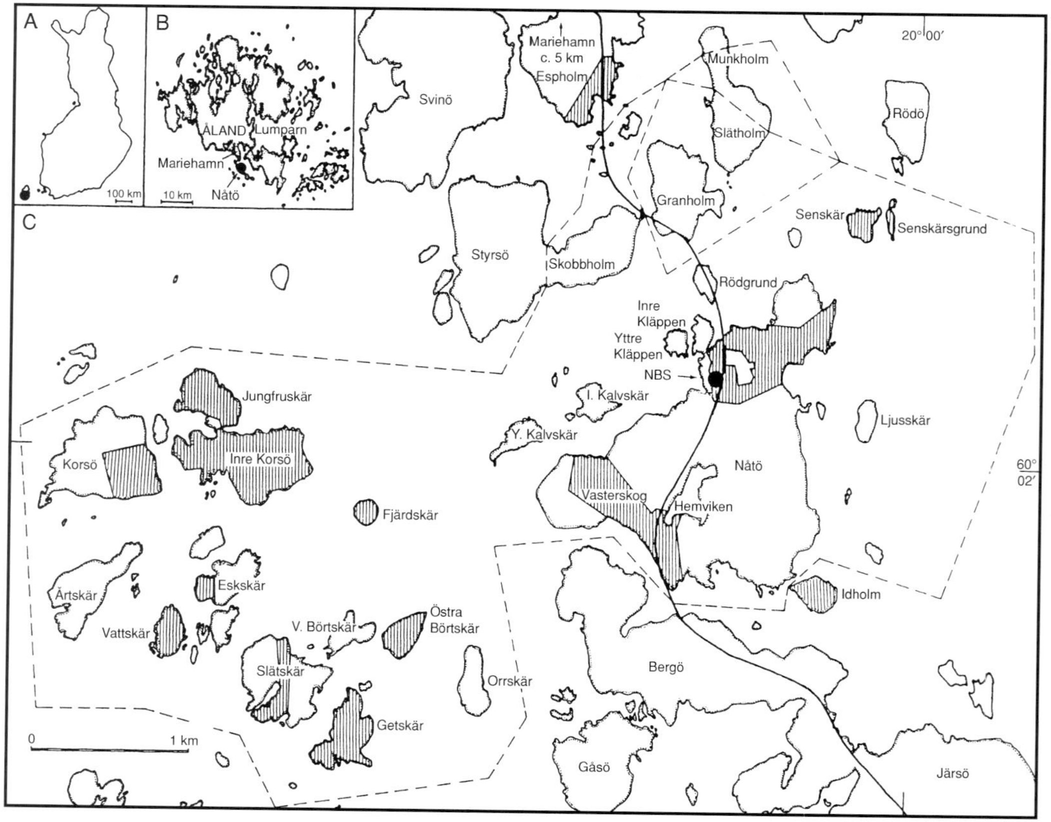

Fig. 6.1. (A) General location of Nåtö at the SW corner of Finland, and (B) location of Nåtö in the archipelago of Åland. (C) Nåtö and its surroundings. The hatched areas belong to the Nåtö–Jungfruskär Nature Reserve and the Espholm Nature Reserve. The hazels studied are all located in the hatched area of northern Nåtö. NBS, Nåtö Biological Station (solid dot).

Fig. 6.2. Hazel stool no. 46 (to the right; basal girth 539 cm, calculated age 620 years according to a land uplift rate of 4.5 mm $year^{-1}$) and stool no. 42 (to the left; basal girth 627 cm, calculated age 722 years). At the far right the stem of an ash pollard can be seen. The vernal aspect is luxuriant comprising among others *Primula veris, Anemone nemorosa, A. ranunculoides* and *Ranunculus ficaria* ssp. *bulbilifer.* (Photo courtesy of C.-A. Hæggström, 31 May 1987.)

The field layer exhibits a striking vernal aspect with many spring-flowering species, several of which are geophytes (Fig. 6.2). Typical species of the vernal aspect are *Adoxa moschatellina, Anemone nemorosa, A. ranunculoides, Carex digitata, Convallaria majalis, Corydalis solida, Dentaria bulbifera, Gagea lutea, Hepatica nobilis, Lathraea squamaria, Lathyrus vernus, Maianthemum bifolium, Oxalis acetosella, Paris quadrifolia, Primula veris, Ranunculus auricomus, R. ficaria* ssp. *bulbilifer, Viola mirabilis, V. reichenbachiana* and *V. riviniana.* A few early flowering orchids, *Dactylorhiza sambucina, Listera ovata* and *Orchis mascula,* are abundant. Later on, other orchids, such as *Dactylorhiza fuchsii, Neottia nidus-avis, Platanthera bifolia* and *P. chlorantha,* are in blossom. Other typical herbs of the field layer are *Allium oleraceum* (chiefly vegetative shoots), *Anthriscus sylvestris, Campanula latifolia, C. persicifolia, C. trachelium, Dryopteris filix-mas, Filipendula ulmaria, F. vulgaris, Fragaria vesca, Geranium sylvaticum, Geum rivale, Heracleum sibiricum, Hypericum maculatum, Laserpitium latifolium, Melampyrum nemorosum, M. pratense, M. sylvaticum, Moehringia trinervia, Polygonatum odoratum, P. multiflorum, Ranunculus acris, Rubus saxatilis, Sanicula europaea, Satureja vulgaris, Veronica chamaedrys, Vicia sepium*

Fig. 6.3. Hazel stool no. 44 (to the right; basal girth 724 cm, calculated age 833 years according to a land uplift rate of 4.5 mm year^{-1}), stool no. 57 (to the left; basal girth 344 cm, calculated age 396 years) and stool no. 49 (to the left of stool no. 57; basal girth 722 cm, calculated age 859 years). Ash pollards can be seen between the hazels. (Photo courtesy of C.-A. Hæggström, 10 April 1998.)

and *V. sylvatica*. Among the graminids *Agrostis capillaris*, *Carex digitata*, *C. pallescens*, *Dactylis glomerata*, *Deschampsia cespitosa*, *Elymus caninus*, *Luzula pilosa*, *Melica nutans*, *Milium effusum*, *Poa nemoralis* and *P. trivialis* are the most abundant.

The ground layer is rather thin and comprises mosses, such as *Cirriphyllum piliferum* (Hedw.) Grout, *Plagiomnium cuspidatum* (Hedw.) T. Kop., *P. undulatum* (Hedw.) T. Kop. and *Rhytidiadelphus triquetrus* (Hedw.) Warnst.

The soil of the hazel stands is of the grey-brown weakly sod-podzolic type (Nykvist, 1968) typical of the most luxuriant stands in south Finland. For detailed soil profiles and study methods, see Hæggström (1983). The humous layer consists of mull-like moder (Kubiëna, 1953). The mineral soil is mostly till with fine sand as the major fraction. The soil is mostly extremely rich in stones and boulders, particularly in localities higher than 5–6 m above the MWL. The pH of the mull-like moder varied between 5.3 and 6.9 (mean 6.1; $n=11$) and of the mineral soil between 5.3 and 7.1 (mean 6.3; $n=13$). The concentration of exchangeable calcium is fairly high, being between 600 and 2122 mg l^{-1} (mean 1299 mg l^{-1}; $n=11$) in the mull-like moder and between 175 and 1065 mg l^{-1} (mean 473 mg l^{-1}; $n=13$) in the mineral soil.

4 Results

The age of the individual hazel stems is usually rather low. The girth of the thickest stem studied was 73 cm and its age was only 50 years. About 15 shrubs had stems with a girth of at least 40 cm. The age of the five oldest stems determined was 60 years (girth 45 cm), 59 years (39 cm), 58 years (36 cm), 52 years (39 cm) and 51 years (36 cm). Still thicker stems were studied on the small wooded meadow island of Senskär in 1977; their stem girth was 88 and 86 cm, respectively, but the age could not be determined due to heartwood rot.

The growth index of the individual stems studied varied between 0.39 and 1.60. The mean growth index of all stems studied was 0.87 ($n = 169$). The basal girth of the 84 stools studied varied between 22 and 861 cm. Altogether, 18 of the studied shrubs had a basal girth over 600 cm. The maximum number of aerial stems of one stool was about 170 (Table 6.1).

The 84 shrubs studied grew at an altitude between 83 cm and 1133 cm above the MWL. The age of these localities as land above the MWL was

Table 6.1. The approximate height, girth and number of aerial stems of the 20 largest hazel stools studied, and the corrected age of the locality (according to a land uplift rate of 5.0 and 4.5 mm year^{-1}) and the calculated age of each stool.

				Age of locality		
Shrub no.	Height (m)	Girth (cm)	Approx. no. of stems	5.0 mm year^{-1} uplift	4.5 mm year^{-1} uplift	Age of stool
54	7	861	50	898	998	990
50	7	831	170	838	931	955
77	9	831	75	1942	2158	955
79	9	827	105	2024	2249	951
75	8	813	95	1908	2120	934
66	8	786	55	1232	1369	903
43	8	782	40	788	876	899
40	7	770	100	780	867	885
61	7	725	50	988	1098	833
44	8	724	120	804	893	832
49	8	722	125	830	922	830
52	7	719	120	878	976	826
45	6	697	85	804	893	801
47	7	675	80	816	907	776
48	9	669	110	830	922	769
53	5	658	90	878	976	756
63	8	653	50	1066	1184	751
42	7	627	80	786	873	721
74	9	594	80	1832	2036	683
55	8	563	40	902	1002	647

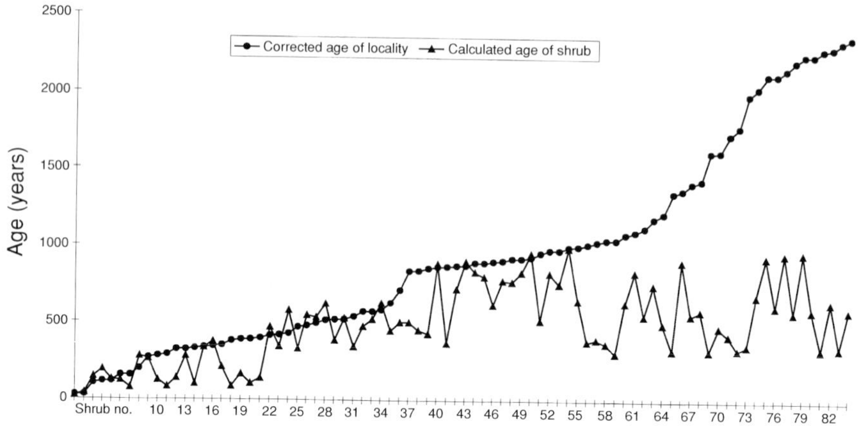

Fig. 6.4. The corrected age of the locality and the calculated age of each hazel stool (according to a land uplift rate of 4.5 mm year^{-1}).

calculated to be between 166 and 2266 years (according to a land uplift rate of 5.0 mm year^{-1}), or 184 and 2518 years (according to a land uplift rate of 4.5 mm year^{-1}). The corrected age of the localities was calculated to be between 26 and 2126 years (according to a land uplift rate of 5.0 mm year^{-1}), or 29 and 2362 years (according to a land uplift rate of 4.5 mm year^{-1}).

The calculated age (according to the mean growth index of 0.87) of the 84 shrubs was between 25 and 990 years (Fig. 6.4 and Table 6.1). Thus the shrubs studied seem to have a definite maximum age of approximately 1000 years. This age is reached when the altitude of the locality is between 5 and 6 m above the MWL. The maximum age of the shrubs growing at higher altitudes was of the same magnitude.

The curves of the calculated age of the shrubs and the curve of the corrected age of the localities follow each other rather well up to about 5 m above the MWL, especially when the land uplift rate of 4.5 mm year^{-1} was used (Fig. 6.4). Due to the methods used, several of the shrubs became older than the calculated age of their locality, particularly if the land uplift rate of 5.0 mm year^{-1} was used in the calculations.

5 Discussion

According to Rackham (1980), hazel stools often reach 6 ft (183 cm) in diameter, which corresponds to about 575 cm in basal girth. Several of the stools on Nåtö are considerably larger. Palmgren (1915–1917) mentions much larger hazels from the Åland Islands. The largest stool, which grew on the island of Granholm north of Nåtö Island, was 18 m in basal girth. On Nåtö Island he reports a stool with a basal girth of 15 m which had about 300 shoots.

A hazel with a thick stem, the girth of which was 103 cm, grew in Björkö Hamnö in Brändö, Åland Islands (Bergroth, 1894). Eklund (1927) reported on seven tree-like hazels growing on the island of Lövskär in Berghamn, Houtskär, in the Archipelago Sea east of Åland; the girth of the two largest was 195 cm in 1924. The thickest hazel tree that has come to my knowledge is reported from Norway (Lagerberg *et al.*, 1952). This tree grew in Yddal in Fusa, Hordaland. Its girth was 446 cm at about 0.5 m height in 1937. At about 1 m it divided into two main branches, the girths of which were 295 and 252 cm, respectively. A thick hazel was reported from Näsum in Scania, southern Sweden; its girth was 219 cm at about 0.5 m (Lagerberg, 1947).

Due to the methods used, some of the shrubs were calculated to be older than the age of their locality, especially when the land uplift rate of 5.0 mm year^{-1} was used in the calculations. This may be due to several sources of error.

1. The mean growth index of 0.87 was used for all stools. Individual growth indices for each stool could give a better result. However, as cores were taken from aerial stems of only 56 of the 84 stools, this was not possible.
2. In the calculations of the age of the stools, it was assumed that the girth increments of the aerial stems and the underground stool are equal. However, this is not certain. As no data are available of the girth increment of subterranean parts of hazels, studies on cores taken from subterranean stools or long-term studies are needed to solve this problem.
3. The age was calculated assuming an even land uplift rate during the last 2500 years. Variations in the land uplift rate may have occurred which could result in wrong ages for the localities for the individual hazels.
4. The exact location of the starting point of a hazel is very difficult to define, due to the uneven bases of the stools in the often boulder-rich terrain.
5. The influence of the boulders on the growth of the hazel stools is difficult, or impossible, to determine. In some cases the stool was found to be partly growing on boulders protruding above the ground. In other cases the stool had divided into smaller separate shrubs. This is particularly the case on higher ground (above 5–6 m) and therefore many rather small stools of today may be parts of previous larger ones.
6. The thickness of the humus layer has evidently increased during the years, which adds some centimetres to the altitude of the locality. This was not taken into consideration.
7. The climate has changed during the last few thousand years. For instance, the Little Ice Age (about 1300–1880 AD) could have hampered the growth. The growth rings also show a considerable variation during the last 60 years. However, narrow or wide growth rings had not formed during the same years in the aerial stems of different stools. Instead, the growth of

individual stems seems to be most dependent on shade from other stems and surrounding trees and shrubs.

8. A hazel which starts its life close to the seashore will be displaced more and more in relation to the seashore due to the land uplift. This means that the shrubs may become more shaded and more sheltered against wind from the cold sea. During the last two millennia, Nåtö has changed enormously. The area of the island has increased, and its position in the archipelago has undergone a change shifting from the outer archipelago zone to the inner archipelago zone (Munsterhjelm, 1997).

9. Man's impact on the hazel stands of Nåtö has obviously been heavy over several hundred years. The oldest evidence for permanent habitation is from 1537 when the three homesteads were inhabited (Hæggström, 1983). The habitation is obviously a few hundred years older. As the hazel stands were used as coppices, at least during the 19th and beginning of the 20th centuries, the growth of the stools has been influenced. It has been proved that ashes that have been regularly pollarded show very narrow annual rings in their wood (Hermelin, 1976; Schweingruber, 1987; Slotte, 1993). Underground stools of hazel are perhaps influenced in the same way. However, aerial hazel stems show a rapid growth with wide annual rings when regularly cut. Most of the hazels near the biological station were cut in 1963 and a few of them later as well. The growth index of 15 of these stools was 0.90 while nine stools of sampling area 8 had a growth index of only 0.73. These latter stools have grown almost without any disturbance since the 1950s. The reason for the faster growth of the cut hazels is that the light conditions are better in sparse stands.

As a result of this study on the age and size of hazel stools, it can be concluded that more long-term studies are needed to produce more accurate data.

References

Bergroth, O. (1894) Anteckningar om vegetationen i gränstrakterna mellan Åland och Åbo-området. *Acta Societatis pro Fauna et Flora Fennica* 11, 1–78.

Eklund, O. (1927) Baumförmige Hasel (*Corylus avellana* L.). *Memoranda Societatis pro Fauna et Flora Fennica* 1, 16–20.

Hæggström, C.-A. (1983) Vegetation and soil of the wooded meadows in Nåtö, Åland. *Acta Botanica Fennica* 120, 1–66.

Hæggström, C.-A. (1988) Protection of wooded meadows in Åland – problems, methods and perspectives. *Oulanka Reports* 8, 88–95.

Hæggström, C.-A. (1996) Nåtö Biological Station. *Memoranda Societatis pro Fauna et Flora Fennica* 72, 227–235.

Hämet-Ahti, L., Suominen, J., Ulvinen, T. and Uotila, P. (1998) *Retkeilykasvio* [*Field Flora of Finland*], 4th edn. Finnish Museum of Natural History, Botanical Museum, Helsinki.

Henriksson, H. (1966) Lövtäkt på Åland. *Åländsk Odling* 27, 94–108.

Hermelin, S.A. (1976) Askstubbens årsringar skildrar bygdens historia. *Lustgården* 57, 38–42.

Kubiëna, W.L. (1953) *Bestimmungsbuch und Systematik der Böden Europas.* Illustriertes Hilfsbuch zur leichten Diagnose und Einordning der wichtigsten europäischen Bodenbildungen unter Berücksichtigung ihrer gebräuchlichsten Synonyme. Ferdinand Enke Verlag, Stuttgart.

Lagerberg, T. (1947) *Vilda växter i Norden II. Corylaceae – Rosaceae*, 2nd edn. Bokförlaget Natur och Kultur, Stockholm.

Lagerberg, T., Holmboe, J. and Nordhagen, R. (1952) *Våre ville planter II.* Revidert og forøkt utgave. Johan Grundt Tanum, Oslo.

Lisitzin, E. (1964) Contribution to the knowledge of land uplift along the Finnish coast. *Fennia* 89, 1–22.

Munsterhjelm, R. (1997) The aquatic macrophyte vegetation of flads and gloes, S coast of Finland. *Acta Botanica Fennica* 157, 1–68.

Nykvist, N. (1968) Brunjord eller 'gräsmarkspodsol?' Några synpunkter på rysk och svensk jordmånsklassifikation [Summary: Brown earth or sod-podzolic soils. Some aspects of Swedish and Russian soil classification]. *Sveriges Skogsvårdsförbunds Tidskrift* 66, 419–426.

Palmgren, A. (1915–1917) Studier öfver löfängsområdena på Åland. Ett bidrag till kännedomen om vegetationen och floran på torr och på frisk kalkhaltig grund. I–III. *Acta Societatis pro Fauna et Flora Fennica* 42, 1–633.

Rackham, O. (1980) *Ancient Woodland – its History, Vegetation and Uses in England.* Edward Arnold, London.

Rackham, O. (1983) *Trees and Woodland in the British Landscape.* J.M. Dent & Sons, London.

Schweingruber, F.H. (1987) Gstumpeti Oesche im Heimisbach. *Schweizerische Zeitschrift für Forstwesen* 138, 449–452.

Slotte, H. (1993) Hamlingsträd på Åland. *Svensk Botanisk Tidskrift* 87, 283–304.

Winterhalter, B., Flodén, T., Ignatius, H., Axberg, S. and Niemistö, L. (1981) Geology of the Baltic Sea. In: Voipio, A. (ed.) *The Baltic Sea.* Elsevier Oceanography Series No. 30. Elsevier, Amsterdam, pp. 1–121.

Witting, R. (1943) *Landhöjningen utmed Baltiska havet under åren 1898–1927.* (Referat: Die Landhebung dem Baltischen Meere entlang in den Jahren 1898–1927.) *Fennia* 68, 1–40.

7

The Study of Charcoal-burning Sites in the Apennine Mountains of Liguria (NW Italy) as a Tool for Forest History

C. Montanari, P. Prono and S. Scipioni

Istituto di Botanica, Università di Genova, Corso Dogali 1c, 16136 Genova, Italy

The study of charcoal as a source of palaeoecological information is a well-known archaeobotanical tool, especially in archaeology. In recent years, beside hearth remains and scattered charcoal fragments or the little-employed soil anthracology, interest has been increasing, especially in France, about the remains of charcoal kilns. Their location can indicate former woodlands which have vanished today as a consequence of long and total exploitation. Moreover, stratigraphic analysis of charcoal-burning sites may sometimes allow researchers to find vegetation successions in connection with forest exploitation. In any case, the identification of the tree taxa employed for charcoal production provides biogeographic, palaeovegetational and palaeoethnological data which can usefully support those coming from archive, archaeological or other palaeobiological sources (e.g. palynology, palaeoecology). In the hinterland of Genoa (Val Vobbia), in the framework of the activities of the Laboratory of Environmental History and Archaeology of the University of Genoa (LASA), research has been undertaken of several charcoal kilns of post-medieval age located between 400 and 600 m above sea level (a.s.l.). First anthracological data are presented, concerning a stratigraphic sequence of kilns buried by colluvium, dating from the 17th century and other ones still in use about 50 years ago. A first provisional picture coming from the anthracological analysis of these sites shows the occurrence of at least 11 taxa: they are mesophilous broadleaves, almost all still widespread nearby today (*Ostrya, Corylus, Quercus* deciduous spp., *Carpinus, Castanea, Fraxinus, Acer, Viburnum, Ulmus, Euonymus and Fagus*).

(eds M. Agnoletti and S. Anderson)

1 Introduction

At the end of the last glacial period, in boreal temperate regions woodland recolonized wide territories, by the spreading of different tree species from their refuge areas. For several thousands of years most of Europe was covered by increasing dense forest vegetation, with different forest climaxes following one another according to variations of climatic conditions. Later on, the spread of man and especially the change from a hunting and collecting economy to a farming and breeding one caused a progressive retreat of tree cover owing to new and more intense methods for natural resources exploitation. Nowadays, we can often see a reversal of this trend, with secondary woodlands recolonizing wide areas in which intensive exploitation has stopped. Both natural and anthropogenic forest dynamics have been studied by several methods and from different points of view in the fields of natural sciences, archaeology, palaeoethnology and history.

The oldest times of forest history (before historical ages) can be studied particularly through plant micro- and macroremains (such as pollen, phytoliths, wood, charcoal, seeds, etc.). In more recent times, many other sources can be added, especially in the historical field (landscape descriptions, agronomy treatises, notorial deeds, historical cartography, historical photographs, etc.).

Among the most useful palaeobotanical evidence for the study of forest dynamics and of forest resources exploitation, is the remains of charred wood. In the last 20 years the study of applied anthracology has been widely developed, collaborating with palynology or sometimes replacing it. Information coming from these two different kinds of approach (i.e. charcoal as macroremains, pollen as microremains) are partly overlapping and partly complementary (Leroi-Gourhan, 1992; Leroyer and Heinz, 1992). In many cases, pollen analysis is devoted to reconstructing the ancient environment of wide areas (regional studies), on the basis of pollen assemblages preserved in lake or bog sediments which collect pollen production and dispersal from areas of many square kilometres. However, pollen assemblages can also provide information of local significance when coming from microenvironments; this is the case of soil profiles, wet soils or small hollows under tree canopies, archaeological contexts or caves. Wood and charred wood remains mainly give this last kind of information, that is a local indication; they are in fact macroremains which are not dispersed long distances from the source. Moreover, wood remains (in temperate climates) are mainly preserved when waterlogged and therefore their preservation is due to particular soil conditions. Charred remains, on the contrary, are well preserved in more or less dry soils, even for long periods (thousands of years) and can therefore constitute useful evidence under the different environmental conditions, from those absolutely natural (e.g. wildfires) to man-controlled ones (e.g. hearths, kilns, charcoal kilns, wood clearing, various kinds of agronomic practices). Each of these different anthropic contexts produces some kind of selection of biotic remains so their significance must be accurately evaluated. In particular, charcoal remains coming from charcoal-burning sites can provide valuable information on technical, trade and economic aspects of a population and on the tree covering of the surroundings; as a rule, indeed, it was not the wood which was transported out of the woodland to be burnt, but the charcoal which was transported to the exploitation site.

Charcoal burning is a very ancient technique and information about it can be found starting from the third millennium BC; its use is archaeologically documented in Egypt and in the near East (Lugli and Pracchia, 1995); in the fourth century BC Theophrastus describes the carbonization method and Plinius the Elder describes plant species suitable for different uses of charcoal and by-products of charcoal burning. A lot of deforestations which were carried out, especially in the Mediterranean basin, are commonly ascribed to the increasing need for charcoal, mainly as a consequence of spreading metallurgy. The same seems to be true also in China, in the Han period (206 BC to AD 220). Sporadic historical news is known for more recent times (the Middle Ages), but it is only in the 16th century that technical descriptions of charcoal burning can be found (Lugli and Pracchia, 1995). This chapter gives a lot of information on this topic concerning mainly central and southern Italy; it also provides lists of plants with the specification of the use of the respective charcoal (mainly metallurgy) and of the by-products of carbonization (wood tar, pitch, potassium carbonate, etc.). The authors obtained many interesting ideas by observing present-day charcoal burners still working in various Italian regions and by other ethnoarchaeological activities.

In Italy, some archaeobotanists (e.g. Castelletti *et al.*, Nisbet, Follieri and Forlani) have been studying for many years the remains of charred wood from various origins, especially in the archaeological field. Nevertheless, the anthracological study of charcoal-burning sites is a rather marginal and recent subject, whereas interest in charcoal kilns from a technological and economic point of view has been held for a long time.

Very ample and precise research on charcoal-burning sites has been in progress for more than 20 years, mainly in France by Vernet, Bonhote, Metailiè, Izard, Fabre, Lejeune and others. Wide documentation is therefore available (Vernet, 1992), especially for the Pyrenees and the French Mediterranean region. The wide experience gained on this topic has made it possible to obtain a well-tested and standardized fieldwork methodology. Moreover, the use of the reflection light microscope to observe directly the fracture surface of the charcoal allows the archaeobotanists to examine a number of fragments in reasonably short times, so that research has acquired a quantitative and statistically valid dimension. The recent book by Vernet (1997) *L'Homme et la Forêt Mediterranéenne de la Prehistoire à Nos Jours* is a good example of how anthracological research, both in the 'traditional' archaeological field and in the field of the environmental archaeology, contributes in clarifying many palaeo-environmental problems, since before history until today (that is historical anthracology and anthracology with a 'high resolution' of human activities, *sensu* Vernet).

2 A case study in progress on the Ligurian Apennines (NW Italy)

For a few years, the ethnoarchaeological study of charcoal-burning sites in the hinterland of Genoa (Ligurian Apennines) has been used as a subject for brief teaching campaigns concerning environmental archaeology for university students. Such activities are part of the aims of the Laboratory of Environmental History and Archaeology (LASA) which has been recently established within the degree course on cultural goods preservation at the University of Genoa.

We will now briefly discuss a few examples of preliminary results coming from field and laboratory work carried out over the last few years. In the surroundings of Vobbia (Scrivia valley) (Fig. 7.1) many charcoal-burning sites have been found, many still in use until 50 years ago (e.g. UT4, Y10, Z12 on Fig. 7.2). In one case, however, a site (UT1) has been excavated, showing a series of different carbonization phases (Fig. 7.3) buried under colluvium;

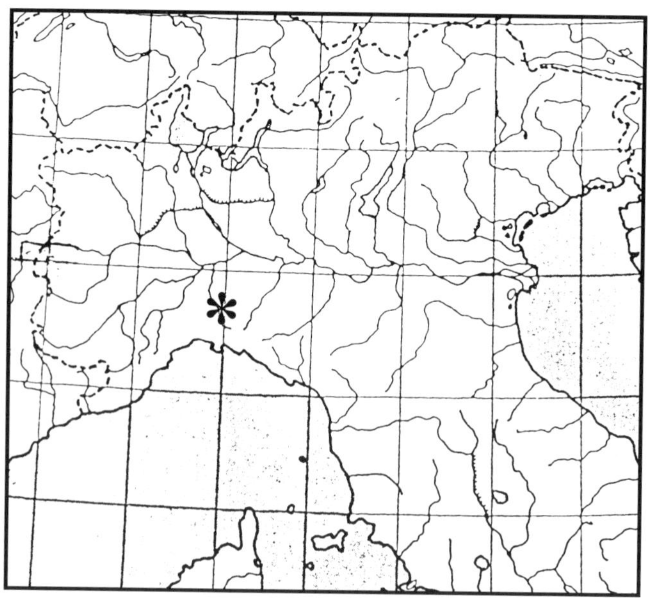

Fig. 7.1. Location of the study area.

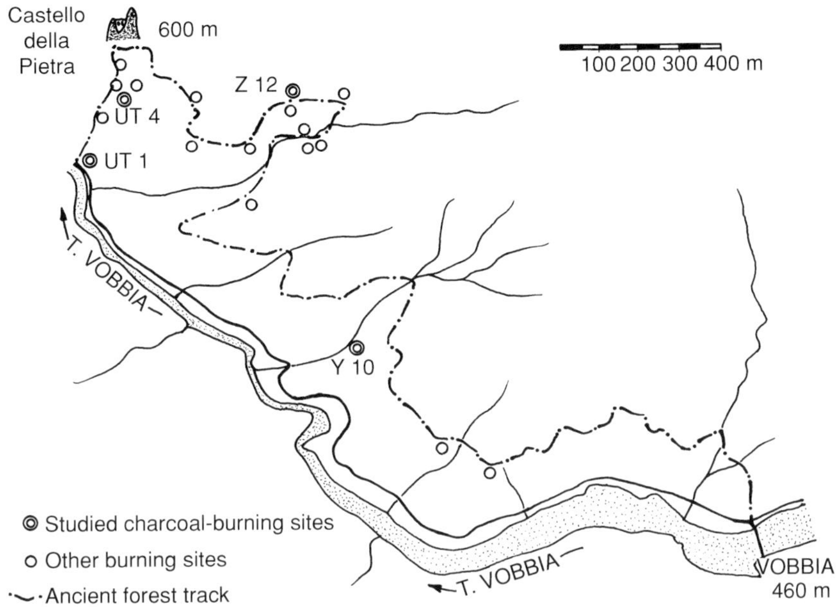

Fig. 7.2. Map of the study area: several charcoal-burning sites are marked, mostly along an ancient forest track linking the village of Vobbia and Castello della Pietra (12th century).

Fig. 7.3. Section of the buried charcoal-burning site UT1, showing two charcoal layers (US6, US9) dating back to the middle of the 17th century.

a radiocarbon dating from the lowermost charcoal layers sets them in the middle of the 17th century (G. Calderoni, University of Rome, personal communication).

The identification of charcoal remains was carried out with the aid of a reflecting light microscope (Nikon Optiphot 100). In some cases, a scanning electron microscope (SEM, Philips 515) was also employed. All the characteristics of the charred wood were observed on fractured surfaces, transversal or longitudinal (tangential and radial). The handbooks of Schweingruber (1990a,b) were used and a reference collection of charred woods as well.

Altogether, at least 11 taxa have been identified. They are all more or less mesophilous deciduous broadleaves, according well with present vegetation. No conifer charcoal was observed and even today conifers (all introduced) are not common in this valley. Pollen analysis of fluviolacustrine sediments in this area, dating back to about 6500 years BP (Montanari *et al.*, 1985), have pointed out the presence of former fir woods which disappeared many centuries ago. The recent charcoal-burning places (i.e. UT4, Z12, Y10; Figs 7.4–7.6) show anthracological spectra more or less similar to the old ones. The anatomical features of the charcoal fragments on which the bark is preserved show that the cutting down was carried out at the winter's end, in accordance with the custom that the last living charcoal burners testify to still today. In both layers of UT1, though with different amounts, hop hornbeam (*Ostrya carpinifolia*) prevails (Fig. 7.7). In the older layer (US9) hazel (*Corylus avellana*) (Fig. 7.8), ash (*Fraxinus ornus*), hornbeam (*Carpinus betulus*) (Fig. 7.9) and elm (*Ulmus campestris*) are also common. The relative abundance of *Viburnum* and the occurrence of beech (*Fagus sylvatica*) and *Euonymus* are worth mentioning. These two last

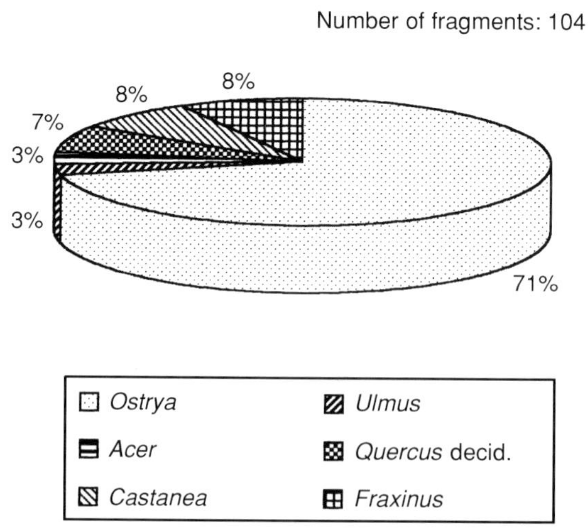

Fig. 7.4. Anthracological spectrum obtained from the charcoal-burning site UT4.

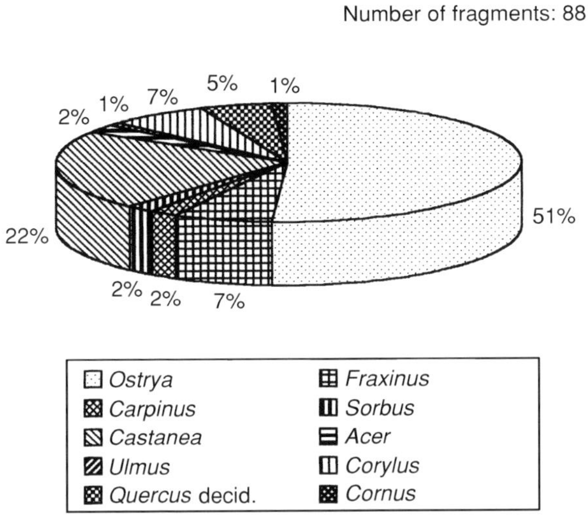

Fig. 7.5. Anthracological spectrum obtained from the charcoal-burning site Z12 B2.

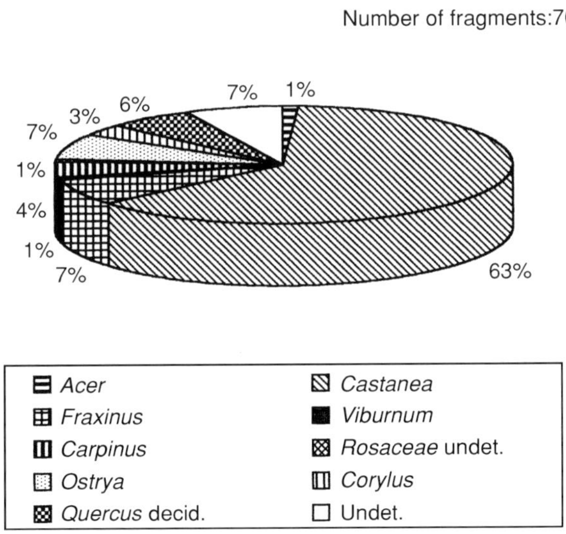

Fig. 7.6. Anthracological spectrum obtained from the charcoal-burning site Y10.

species have not been found in the more recent layer, US6, of the same section or in recent burning sites (20th century). At present, beech seldom occurs at such low altitudes (400–500 m a.s.l.), while scattered bushes of *Viburnum lantana* and *Euonymus europaeus* seldom grow in the underwood. On the whole, the layer US9 shows a good floristic diversity and, from a quantitative

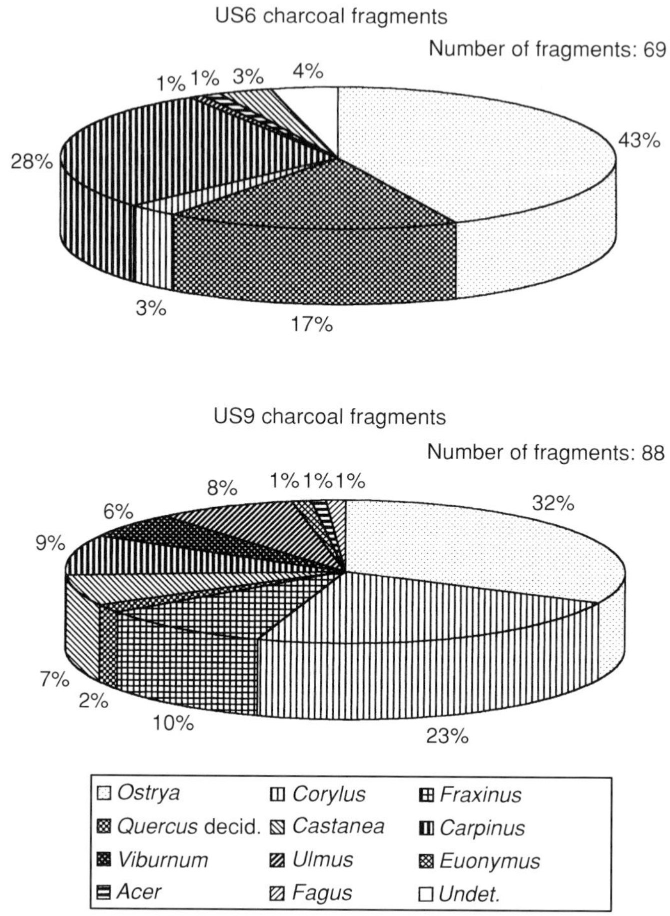

Fig. 7.7. Anthracological spectra obtained from the charcoal-burning site UT1.

point of view, a wider sharing among different taxa. The more recent layer (US6) shows less taxa, with hop hornbeam, hornbeam and deciduous oaks widely prevailing. A clear predominance of a single taxon is shown particularly in recent charcoal-burning sites, with a maximum in UT4 (*Ostrya* = 71%, Fig. 7.4). Nevertheless, sites Z12 B2 and Y10 still show a good floristic variety (ten different taxa, Figs 7.5 and 7.6).

Therefore, the surrounding woods seem to have maintained a relative uniformity for a long time, even though they have been intensively exploited. The usual practice must have always been to cut and burn every woody plant growing in a selected area, shrubs included, to make charcoal. Our anthracological analysis, indeed, shows twigs of small size prevailing (diameter 1–5 cm) which can be due to a greater care of burners in collecting

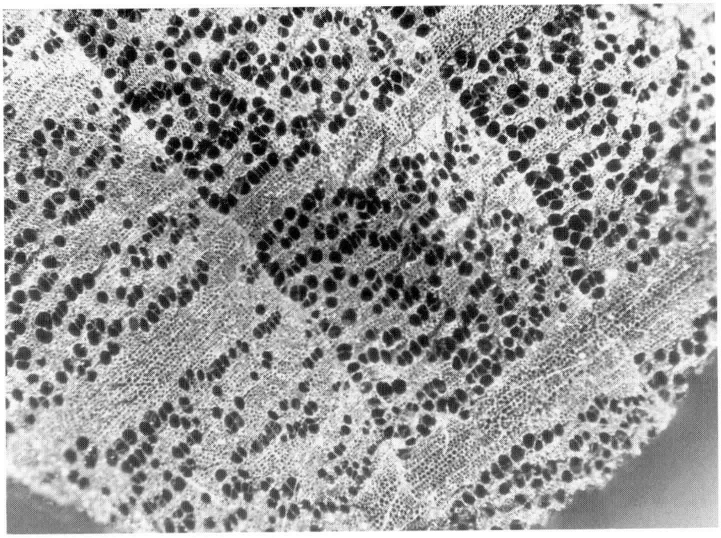

Fig. 7.8. Transversal section of charcoal from *Corylus avellana* (reflecting light microscope, ×50).

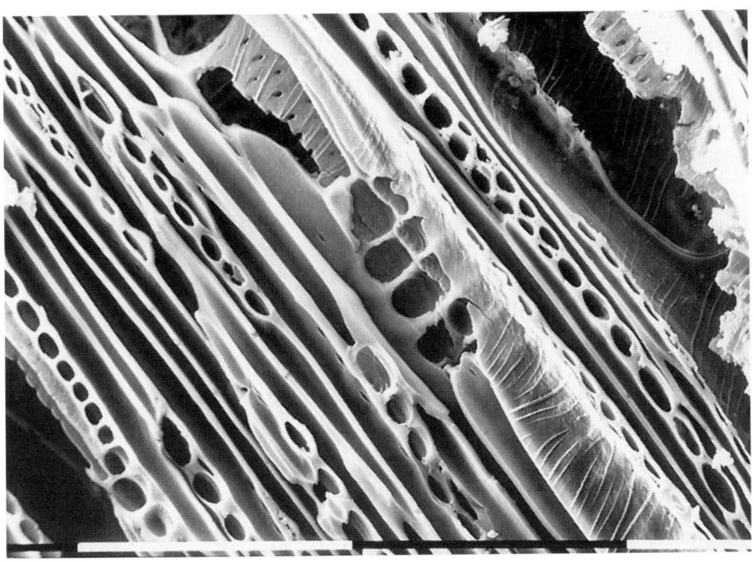

Fig. 7.9. Longitudinal (tangential) section of charcoal from *Carpinus betulus* (SEM, scale bar = 0.1 mm).

larger, more valuable, charcoal fragments or to a very short cutting interval which was employed in the past, when charcoal production was one of the most important activities in this area. The use of chestnut too is worth some remark: if the information coming from the buried charcoal-burning sites is representative of the charcoal production standards of previous centuries, it is clear that chestnut was scarcely employed for charcoal because of the much higher value of its fruit. However, in the more recent charcoal-burning sites, high percentages of chestnut are found (e.g. Y10 and Z12 B2), pointing out this species crisis in connection both to the phytopathologies it undergoes and to the decline of the economic value of its fruit.

In the particular case of the Z12 B2 burning site we have compared recent vegetation with the results of our anthracological analysis. As a result *Acer, Carpinus, Castanea* and *Corylus* have been found both as charcoal and in recent vegetation, whereas *Fraxinus, Ostrya, Quercus* (deciduous spp.) *Sorbus* and *Ulmus* have been found only as charcoal remains on the charcoal-burning site, but not in the recent surrounding vegetation.

These first results also attest to the forestry importance of hop hornbeam in the last centuries: nevertheless, the Holocene spreading of this species still needs to be clarified in many respects. In pollen diagrams it seems to have never been as important in forest composition as it is at present. This could be connected to its use as fodder (pollarding, coppicing and using the suckers as fodder) or with a recent (post-medieval) spread.

3 Conclusion

The interest of studying charcoal remains from charcoal-burning sites as a source of information for forest history is widely established. Charcoal production has been a widespread activity since very ancient times and its methods are rather simple and homogeneous (at least with regard to the evidences they leave behind), so this field of research is of great interest in reconstructing forest history, both for studies of local history and for wide regions. Obviously, as these data have an anthropogenic source, they must be evaluated in a man-controlled context; nevertheless, very useful information can be obtained even with regard to the natural environment, for instance from a floristic point of view (biodiversity) and with respect to the history of the upper tree line. What we present here is only a small example, of course, but many other facts can be derived from charcoal spectra or even from the occurrence of charcoal-burning sites themselves – for instance, hazel charcoal was in demand for gunpowder (D. Moreno, personal communication); Vernet (1997) has shown the occurrence of charcoal-burning sites over the present tree line in the Pyrenees and in Mediterranean France (as a clear evidence of deforestation). The same is recorded by Alessandrini (1984) on the Emilian-Ligurian Apennines.

Today an experimental charcoal kiln is burning near Vobbia, in order to record each phase of the manufacturing process, to try to understand the functional, physical and chemical characteristics and the effects on the site (Figs 7.10–7.12). Another charcoal kiln has been recently burnt on the Portofino promontory, as a demonstration.

Fig. 7.10. Experimental charcoal kiln in Vobbia: the pile.

Fig. 7.11. Experimental charcoal kiln in Vobbia: burning phase.

Fig. 7.12. Experimental charcoal kiln in Vobbia, final phase: the charcoal-burning site after the charcoal has been removed.

Acknowledgements

The authors are grateful to teaching staff (particularly Professors D. Moreno, G. Leonardi and G. Calderoni) and students of the three environmental campaigns in 1996, 1997 and 1998 to Vobbia, and to the personnel of LASA who allowed this research to be undertaken.

References

Alessandrini, A. (1984) Il Pino mugo. In: AAVV *Il Monte Nero.* Regione Emilia Romagna, p. 110.

Leroi-Gourhan, A. (1992) Wood and pollens: complementary studies. In: Vernet, J.L. (ed.) *Les charbons de bois, les anciens écosistemes et le role de l'homme. Bulletin de la Societé Botanique de* France No. 139. *Actualités Botaniques* 2–4, 273–280.

Leroyer, C. and Heinz, C. (1992) Complementary in palynological and anthracological studies: pyrenean examples of La Balma Margineda (Andorra) and Belesta (P.O. France). In: Vernet, J.L. (ed.) *Les charbons de bois, les anciens écosistemes et le role de l'homme. Bulletin de la Societé Botanique de* France No. 139. *Actualités Botaniques* 2–4, 281–295.

Lugli, F. and Pracchia, S. (1995) Modelli e finalità nello studio della produzione di carbone di legna in archeologia. In: *Origini Preistoria e protostoria delle civiltà antiche*, Vol. XVIII. Bonsignore Editore, pp. 425–479.

Montanari, C., Guido, M. and Petrillo, M. (1985) Tracce di un bosco di Abete bianco in Val Vobbia messe in luce dall'analisi pollinica. *Archivio Botanico e Biogeografico Italiano* 61, 169–184.

Schweingruber, F.H. (1990a) *Anatomy of European Woods*. Haupt, Bern.

Schweingruber, F.H. (1990b) *Microscopic Wood Anatomy*. Swiss Federal Institute for Forest, Snow and Landscape Research, Birmendsory.

Vernet, J.L. (1992) Les charbons de bois, les anciens écosistemes et le role de l'homme, *Bulletin de la Societé Botanique de France* No. 139. *Actualités Botaniques* 2–4.

Vernet, J.L. (1997) *L'homme et la fôret mediterranéenne de la Prehistoire à nos jours*. Editions Errance, Paris.

8

Local Economic History, Environmental History and Forest History: Some Swedish Experiences and Suggestions

S. Gaunitz

Department of Economic History, Umeå University, S-90187 Umeå, Sweden

Economic history and forest history are two multidisciplinary fields which combine social science approaches with studies of resource use and environmental changes. Economic history can contribute knowledge about social and institutional arrangements, and forest history can add technological and ecological aspects. Local and comparative studies are recommended in order to reach a full understanding of the interdependence between economic, social and ecological factors affecting forestry and forest development. They will also contribute to methodological understanding and improvement. Regional studies of the impact of industrialization upon regional forest development and changes in timber transport systems are recommended for further research.

1 Introduction

There is an urgent need for economic historians to clarify their relations to environmental history and vice versa. An important issue to deal with in this realm is forestry and forest history. The first impression of forest history may be that it should be thought of as part of environmental or ecological history or maybe only botanical development history. However, like economic history, forest history is multidisciplinary, and a closer look will reveal that it often deals with technological, economic and social issues. Thus a multidisciplinary approach is already practised, where social sciences, natural sciences and history – especially its social science branch, economic history – can meet and profit from each other's knowledge and experiences. It is argued here that it would be valuable to forestry and forest history if economic historical theories and methods were used together with the much more commonly used background information about historical conditions. In the future, hopefully, this integration with economic history will contribute even more to a fuller understanding of forest historical and more general environmental and ecological issues, as both approaches are dealing with the same broad contexts and very long time spans.

In an attempt to clarify the opportunities for economic historical research to contribute to the discipline of forestry, this chapter will begin with some aspects taken from a study area which can be labelled as 'local economic history as a means of understanding ecological change'. The word 'ecological' is used in order to underline the importance of the biological aspects of the environment. The local or small-scale research field has been chosen because it often gives substantiality to the issues risen by the scientists and makes it possible to bring them closer to theories of economic history in discussions and understanding.

As usual in economic historical resource analysis the point of departure is the Malthusian and Boserupian approaches to the population problem. However, a closer look will reveal that they come to a halt in front of ecological issues, and for that reason natural scientist knowledge and methods have to be added. The approach which is suggested here starts with the notion of a relationship between man and nature in a local context, or to put it in other words, a study of social and economic man in his and her interdependence with biological nature.[1] The key word here is interdependence, and it is essential to notice that this means that a dynamic approach is chosen. Any evidence we have suggests that both ends of this relation are changing very rapidly, and a static theoretical approach would be a severe hindrance to our understanding. Economic history provides a theoretical framework for how we can observe and interpret changes in human societies: changes through settlements, urbanization, new lifestyles and new demands. Ecological reception of these social changes varies from exhaustion of easily extractable resources to overcrowding of weeds, mice and seagulls.[2] Some of these responses impact directly upon man, but more often the satisfaction of human demand creates imbalances between man and nature which in the long run will impact negatively upon both parties.

The usual response of humans to such imbalances has been to try to extend their control of human society into new realms of nature through new sets of rules, thereby transferring nature into natural resources. The characteristic feature of natural resources is that they can be controlled by members of human society by means of ownership, leases or whatever is the legal or customary fashion of the time. When this has been done nothing but the user's evaluation of the resource has to be taken into account and all the rest of the natural object in question will go unobserved. This can probably best be exemplified by a traditional countryside setting, where the forests and sometimes the grassland of the villagers were regarded and treated as commons whereas ploughland, pasture and gardens were divided amongst individual 'holders'.

Ownership means control of resources, and the study of this in the form of property rights is of prime interest in modern economic history, especially through the work by Douglass C. North and a lot of followers.[3] They regard property rights as a basic institution in our society and stress the need for studies of institutional history. North's most famous work is his application of this approach to macro-explanations of the dynamic changes of economic systems over centuries, but it has also been applied to local issues and special cases.

There is a need for comments on the concepts of the local economic historical approach referred to above: how can local economic history be used as a means of understanding ecological change? The answers may be as follows:

1. Economic history deals with the balance between man and resources or, more accurately, the man and his resources. Generally much attention has been given to the balance of macro-variables in a national context – leading to a scenario of exhausted world resources or global overcrowding. In the local scene, the focus remains on the balance of resources, but the issues may be dealt with in a much more comprehensive way, and they can easily be related to social life and material resources – to trees, forests, agricultural land, river beds and pit holes. Still, it is the human aspects of these 'resources' we are dealing with.
2. Understanding may be interpreted as being aimed at the integration of another, maybe a new or a formerly unrecognized, field in our common body of knowledge. To include ecological change into economics is to broaden the view; the same is true of the inclusion of change – or history! – into traditional neoclassical economic analysis. To introduce changes of economic institutions, such as property rights, into an analysis of the changes of forestry practices will also add new perspectives.
3. Ecological change deals in this context with the changes that take place within the biosphere and which are sometimes caused by human activities, created by our society, and sometimes by the intrinsic development of plant or animal species, and sometimes by physical changes or obstacles.

There is a conceptual issue to be observed here: the biosphere may or may not be included in the resource base of human society, but what we observe is always related to our society and thereby, as such, is part of our resource base. In this sense, environmental history is always a social science.

Finally, a comment on two important concepts in historical perspective: time and comprehensiveness. As historical disciplines, both economic history and forest history relate themselves to the concept of time. In the case of social sciences time often means 'model time', whereas economic historians prefer to talk about 'historical time' and use the order of occurrence in causal explanations. Thus time explains both casual relationships and social settings in economic historical contexts.[4] The same ambiguity can be found between forestry and forest history, and it is important to notice that historical time should be preferred in dynamic approaches. This is also an important difference when it comes to 'semihistorical' disciplines with a preference for model construction such as social anthropology or ethnology.

In economic history there is a tendency towards comprehensiveness or totality in the search for causal explanations and descriptions. It can be attributed to the French tradition of 'histoire totale', and it may be particularly pronounced in certain fields like worldwide or local history, but it is also influential in many other economic historical interpretations.[5]

2 Forest history and economic history

The focus in the following paragraphs will be upon the social and economic aspects of forest history and the human impact upon forests. Crossing Western Europe by plane you might simply think that they have disappeared – making forest history a kind of landscape archaeology. But this is not the real story, and certainly not the whole story. The forests have given room for a more efficient land use. The term 'efficient use' will include a European model of regularly and intensively practised forestry in the central areas of the continent and the import of forest products from less valuable land at some distance, for instance from Sweden. Clearly this way of reasoning exemplifies the need for a broad understanding of the social and economic relations in any general approach to forestry and forest history. It is a macro-approach which applies the ideas of interdependence and totality. It is suggestive of a debate about the importance of climatic changes in Europe during the medieval and the early modern ages. It was some time before the debaters concluded that two or three degrees of sinking temperature in Europe could have different effects on economic performance in the north to those in the south. Environmental changes clearly may have contributed to the changes in the balance of power between nations and regions.[6]

Economic history has traditionally dealt with issues of resource use and especially with the contribution of natural resources to economic growth and development. In ancient and early modern history the supply of food, fuel and timber has been used as an integral part of the explanation of the upswing and downturn of empires. Additionally, the control of basic resources has been a very important issue in the new but also very old field of institutional development and institutional economics. The above-mentioned Douglass C. North is the best known spokesman for this approach.[7] When economic and social historians have studied the transitional periods of peasant societies, which have changed either from feudal practices to market orientation, or from traditional management of farming and forestry to industrial organization and practices, explanations have been sought for in local contexts. Thus we arrive at a field of local history aimed at generalizations. One example of this is the debate about 'the tragedy of the commons', which in the economic historical context has focused upon the supposed expulsion of tenants and cottagers from manorial villages by means of expropriation of their customary property rights on the commons during enclosure movements. Basically this is an application of economists' theory of the prisoners' dilemma, which more or less is static and instrumental to long-term historical processes of change. In genuine forest history the emphasis would be upon various uses of the common land, either for a nature- and biodiversity-oriented purpose or for plantation forestry or, of course, for the conversion of forest land into pasture and arable land. This may be a good example of the difference between, on the one hand, forestry and forestry engineering, which is instrumental and aims at the best economic or biological use of forest land, concentrating upon practical research and education of foresters; and on the other hand, there is forest history, which can be seen as an explanatory natural and social science, aiming at the understanding of forest development and changes and thus becoming an essential part of environmental history.

3 Timing and consequences: economic and environmental life cycles

One of the specialities of economic history is social demography and population development. Migration movements have often gone into resource-rich areas, for instance forests and river basins, reached a population peak with a rich social and cultural life in the resource-rich countryside and then returned to a reverse movement back into urban areas, emptying the resource-rich land of people. In northern Sweden with its late colonization we can easily observe this life-cycle pattern of the settlements; their population development is closely related to the exploitation of forests and prospective farm land. In the first phase we find population growth in scattered agricultural settlements, followed by a build-up of social functions such as trade, schools, churches and other institutions in lots of small centres. Then again there comes depopulation of those settlements and small centres and concentration to a few somewhat larger and totally dominating local towns with urbanized life styles. In the last few decades, a new social and economic system has made it possible to continue to extract resources from the land by an urbanized population living in the new rather distant centres, up to 50–100 km from the forests. Thus these areas went from sparsely inhabited wilderness to human-dominated countryside and then back into human emptiness but not wilderness during the course of about two centuries, from the 1740s to the 1970s. The term 'wilderness' is here used for some kind of natural or virgin habitat.

To forest historians *per se* this economic historical pattern means that the life cycle of local societies will coincide with the new establishment of the forest service and the changes in its institutional arrangements as well as with changing forest practices and, of course, many aspects of biodiversity. The concept of sustainability has been applied both to the forests themselves – in an ecological sense – and to the villages, which have based their economy upon these forests. The importance of changes of economic policy for forest practice can be exemplified with the following example from northern Sweden. In the 'good old days' with sparse population, some 200 years ago, the forest was left almost by itself with few rules, regulations and customs to be observed. There followed a period of intensive colonization and also timber cutting, which led to a build-up of local forest services and minimum dimension laws for cutting in order to protect timber growth. Today it will probably be managed with less forced control but within a very sophisticated set of new rules, including such large-scale market-oriented regulations as the Forest Stewardship Council certification system recently introduced in Sweden.[8] Thus northern Swedish forest practices have had to adapt themselves from pretty much free forest use by local parties to legislative restrictions to international consumer market interference.[9]

Another illustrative example where economic history can contribute to forest history is the regional timing of the industrialization process. Depending upon at which stage of global industrialization the region was drawn into it, the impact upon forest ecological conditions differed. This may be exemplified from the Swedish case. It is obvious that the availability and the quality of forest resources contributed decisively to the settlement expansion and population increase in Sweden during the 200 years since the industrial revolution first started in Great Britain. While the southern and central Swedish plains have a very long colonization history, going back to the Iron Age and Middle Ages, the more peripheral parts of the country were settled rather late by domiciled peasants. In the southern highlands the peak turning point – as measured by the population – came as early as the 1850s whereas this happened almost 100 years later in the northern forest areas. To economic and forest historians alike it is important to find that in the southern forest areas the first heavy exploitation of its timber resources started during a 'proto-industrial' era before any industrial take-off.[10] In addition to an increasing use of wood for farming purposes, some traditional manufacturing industries, mainly iron and glass manufacturing, depended heavily upon wood as fuel from forests at rather short distances. Thus these proto-industrial mills and their managers contributed to the shaping of the early local practices of forest exploitation and resource development. In contrast, the northern forests were opened up much later, at the time of the industrial breakthrough and the accompanying urbanization in Western Europe, and their timber was exported as semimanufactured goods, either hand- or mill-manufactured. At this time the influences of the foreign markets went directly into the forest and the forestry! Obviously the social impact differed between these two periods and areas. This can be described in terms of technological lags and differences in the strength of the links between demand and supply regions at different years, and especially in the year-by-year increasing volumes of the flows of goods going south-west to the urban centres of the western world.

4 Scarcities, relative prices and marginal land

Another argument for the use of economic historical methods in environmental history is our attempt to find explanations of why changes in land use have occurred and will continue to occur in the future. In the study of the interchange between grassland, arable land, forest land and urban plantations and housing, traditional economic methods as well as institutional economic history will be essential; in particular, through examining relative price analysis one might find subtle expressions of man balancing his exchange with nature.

Some of the most well-known studies of forest resource scarcity and transfers of technology into new regions deal with the timber crisis in England in the 17th century and the following development of iron works and canon foundries in Sweden and, half a century later, timber industries in Norway. Several researchers, among them Carlo Cipolla, have presented evidence of this development by use of relative prices. The real meaning of this scarcity – does it mean waste land, or rather better and more profitable market opportunities for timber? – has been much discussed.[11] Another attempt to use relative prices in what may be called an environmental history approach has been made by Nils-Gustav Lundgren, who tried to describe the resource use in the Lule river valley in northernmost Sweden. He has traced the beginning, the peaks and downturns of the various natural resources that have been in prime use by the local population since medieval times: furs, salmon, reindeer meat and wood products like tar, potash and timber. An increase in relative prices is interpreted as evidence of high demand and increased market pressure on the resource.[12]

Lundgren's study was produced in an interdisciplinary research project about the Lule river valley from archaeological times to the 1980s. It included archaeology, ethnology, Scandinavian languages, geography and ecological forest botany together with history and economic history. The results are presented in a very interesting summing-up publication, in Swedish only.[13] From the viewpoint of this chapter, it is regrettable that the participants specialized in their own favourite time spans. Archaeology and forest ecology went as far back in time as they could, unlike the economic historians, who chose to discuss the period after 1850 in the inland areas of the more than 400 km-long valley. Thus the opportunity for interdisciplinary cooperation in a local historical context was not really tested. But some comments may be made.

For one thing, it is obvious from ecological botany studies that the land use in this northern area has changed over time: it is shown, for example, that what was once ploughland is now sometimes recognized as 'virgin forest', and vice versa.[14] This is a kind of circular dynamic which historians must be aware of in their treatment of ecological issues, if not in most causal analyses. A methodological observation is that the difference in sources for each research field and the methods to 'read' them make conclusive common results difficult. In a way both ecologically and economically oriented researchers aim at generalized explanations, but the 'totalities' which they see are not the same, and the coordination of readings, experiments and interpretation becomes a trial to their patience.

5 River valley studies: regional case studies

In the reports from the Lule River Project there are some good examples showing opportunities for economic–historical contributions to a common approach with forest historians in recent years. The reports concentrate on the local development of the inland area of the Lule valley in the period 1870–1970. In particular, they look into how the population supports itself from the local resources and how this support system is transformed by population increases, technological changes and changes in the availability of new natural resources – agricultural land, timber trees, timber land and waterfalls. It is a clear case of a life-cycle pattern based upon local resources, which started with export-oriented growth (timber and later hydroelectric power), feeding an increasing population. The cycle ended with a decline in the number of inhabitants following the exhaustion of prime resources and an immense improvement in technology and increase in productivity and monetary income per capita.[15]

These studies concentrate upon the social and economic aspects and leave out the impact upon the environment. They also deal with how large land areas were used, how much timber was cut and how large were the investments made in floating channels, and also how much labour was required for resource exploitation and earning a decent customary livelihood. They look at our questions from one angle: What did the natural resources do for the population? The other side is left out: What impact did the people have upon nature?[16]

6 River valley studies: local history and case studies

What could be done if forest historians and economic historians chose to use written sources in a local study of the last 200 years or so, that is to say the period of European industrialization? It could, preferably, be done in a well-defined area, for example a river valley similar to the above-mentioned Lule valley or on an island. It would also be interesting to compare northern and southern valleys, such as the northern Vindel river near Umeå and the south Swedish Em river (Vindelälven and Emån). There are lots of written sources dealing with taxation, ownership of land, land surveys, forest service documents and individual farmsteads and manors and, in the north, land-owning forest companies. Ditching, forest road building and reforestation have also been reported as they were subsidized by the government. Thus most documents are of economic or legal origin. An illustrative example of this is that some of the best statistical information about wildlife in the 19th century is given by the bounty paid for beasts and birds of prey, as reported by the state forestry officers. In private archives one can find timber journals and floating reports which have figures about almost all the timber cut in the forests and sawn at the mills for the period of the 1880s to the 1940s in many north Swedish river valleys. Sometimes the size of the timber, its species, its value and the forest district where it was cut can be found in the books. Reports by the forest service and personal letters deal mainly with forest work and the condition of the forests and the opportunities for handling forestry according to formal instructions and practices. Thus the timber and its wood content is focused on, aspects of biodiversity are reported less frequently. A schematic but valuable source is the classifications found in the land surveys and forestry planning surveys, which have mainly been done during the 20th century.

7 A case for comparative approaches: timber transportation systems

Sometimes exploitation requires very large investments and a major interference in the social and economic pattern of a region. Economic historians prefer to talk about development blocks.[17] These blocks often start with the introduction of a basic innovation, for example new technology for timber-floating on a large scale, and this is spread throughout the whole region. The spokesmen for this block will have decisive power over the economic, social and environmental development of the region, including investments, property rights adjustments and forestry practices. Some of the most intriguing research issues will be found in periods of stagnation and downturn. Timber-floating in the northern rivers can be used as an example. In Sweden of the 1960s and 1970s floating was regarded as a rather special or even odd transport system which could easily be abandoned. As historians we want to know why this extensive net of waterways was closed down, and in what interest it was done. It is not only a matter of how it was done and how it functioned, which is so often described in forest and local history literature! In the analysis attention should be paid to the whole life cycle of the floating and also to alternative systems in use or considered. The build-up of this transport development block in northern Sweden began at the beginning of the 19th century, when the state introduced waterway construction for floating on a large scale. Later on private firms took over, but the state once again intervened in the 1870s and 1880s in order to make them publicly available for every competing firm. This was successful until the 1950s, but with increasing concentration among forest companies the few which were still in action were responsible for the closing down of the system in the 1960s and 1970s. The timber transports were finally taken over by lorries after a 50-year period of competition from the 1920s onwards. It is possible to point at some purely economic and technological motives for the closure. By use of land transport the time span between timber cutting and delivery at the mills could be decreased by months, even up to a year, and this led to two very important changes in the whole forest industrial system. The first one was the savings in interest cost and the second the opportunities to handle fresh timber very soon after felling in the production processes. In the beginning this change caused technical problems, but it also gave an impetus to new production methods and higher qualities.

After the closure of the timber-floating there was a more than 10-year-long period of restoration of the rivers, so that their natural condition should be reconstructed, including their function as salmon reproduction areas. The encroachment had increased with each new innovation in floating and digging technology – as it proceeded from spade to bulldozer – and often the former natural rivers had become simply floating channels.

Thus the economic and ecological aspects of the development block, in this case a timber transport block using river floating technology, should be analysed together.[18]

In other parts of Europe, for example in northern Italy, the rivers as a means of transport were replaced by a network of railroads in the latter part of the 19th century, and this in turn was taken over by roads and lorries in the 1940s or 1950s. Probably many of these early railroads were constructed for the only purpose of extracting timber from the woodlands, but some of them became part of a more efficient transport system which could be used in an integrated regional network. Contrary to the floating channels they could be used for two-way traffic and formed a basis of the infrastructure required in an industrialized society.[19]

I conclude with a recommendation for an economic historical approach to forest history, and the above-mentioned comparison between various timber transport systems can be used as an illustration of what can be achieved by means of cooperative research in the two fields. The study of institutional arrangements and institutional change is one of the main building stones of economic history and obviously it has important implications for forest management and forest ecology over time. A suggestion for a comparative approach could be to study the move from water transport of timber to railroad transport in the late 19th century in Italy, Austria or Germany together with the Swedish closure of its waterways and transfer to a road transport system 100 years later. Such an approach would certainly present much knowledge about the institutional and organizational conditions for decisionmaking and change, and we would also know much more about how individuals and local groups react to basic changes in society. We would also learn about the interplay between technological changes and social organization. However, the most important gain could come from our understanding of how nature and society affect each other – that is to say, our understanding of the interaction between economic development and environmental change.

Notes

1 In addition to Esther Boserup's (1965) classic work *The Conditions of Agricultural Growth* (Earthscan, London, 1993) mention must be made of Wilkinson, R.G. (1973) *Poverty and Progress: an Ecological Model of Economic Development* (Methuen, London).

2 One of the first to introduce this broad environmental perspective in a global horizon was Crosby, A.W. (1986) *Ecological Imperialism. The Biological Expansion of Europe 900–1900* (Cambridge University Press, Cambridge).

3 A seminal work in this field is North, D.C. (1990) *Institutions, Institutional Change and Economic Performance* (Cambridge University Press, Cambridge). A case study of methodological interest, though like North's work not explicitly dealing with forestry, is Ellickson, R.C. (1991) *Order Without Law. How Neighbours Settle Disputes* (Harvard University Press, Cambridge, Massachusetts).

4 Bäcklund, D. (1996) Jord, skog och vattenkraft. Ekonomisk omvandling norrländsk skogsbygd 1870–1970 [Land, forest and water power. Economic transition in north Swedish forest areas 1870–1970]. In: Baudou, E. (ed.) *Att leva vid älven. Åtta forskare om människor och resurser i Lule älvdal* [To Live on the River Side. Eight Researches about People and Resources in the Lule River Valley]. CEWE Publishers, Bjästa, pp. 223–249.

5 The Annales School started in the 1920s and has published one of the most famous journals in the field, *Annales.*

6 Utterström, G. (1955) Climatic fluctuations and population problems in early modern times. *Scandinavian Economic History Review,* 3, 3–31. An introduction to the economic historical debate is found in Anderson, J.L. (1981) History and climate: some economic models. In: Wigley, T.M.L. (ed.) *Climate and History. Studies in Past Climates and their Impact on Man.* Cambridge University Press, Cambridge, pp. 33–355.

7 North, D.C. (1981) *Structure and Change in Economic History.* W.W. Norton, See also, for example, Ostrom, E. (1990) *Governing the Commons: the Evolution of Institutions for Collective Action.* Cambridge University Press, Cambridge.

8 The Forest Stewardship Council (FSC) is an international organization for the creation and supervision of national standards of forestry which forest owners can agree to follow, thereby achieving a standard certification of good management of their forests, to be used as a sales argument in the market.

9 Similar examples can be found all over Sweden. However, they differ both in historical timing and in the diversity of the forest stands and also in land ownership.

fore it is necessary to attribute this example to the northern pine and spruce-dominated forest on the western side of the Gulf of Bothnia.

10 The economic historians' proto-industrialization debate was very intense some decades ago. It dealt with the transition stage from merchant capitalism to industrial capitalism, and even though it focused on the transfer from handicrafts to industrial work and the capital structures behind this process it also considered the use of raw materials and their availability.

11 Cipolla, C.M. (1965) *Guns and Sails in the Early Phase of European Expansion 1400–1700*. Collins, London.

12 Lundgren, N.-G. (1987) *Kampen för naturresurserna. Ekonomisk utveckling och institutionella förändringar i Lule älvdal under 700 år [The struggle for natural resources. Economic development and institutional change in the Lule river valley during 700 years]*. SNS, Stockholm.

13 Baudou, E. (ed.) (1996) *Att leva vid älven. Åtta forskare om människor och resurser i Lule älvdal [To Live on the River Side. Eight Researches about People and Resources in the Lule RiverValley]*. CEWE Publishers, Bjästa.

14 Segerström, U. (1996) *Naturmiljön, agrikulturen och människans påverkan på vegetationen i norra Norrland [Natural environment, agriculture and influences of man upon the vegetation of northern Sweden]*. In: Baudou, E. (ed.) *Att leva vid älven. Åtta forskare om människor och resurser i Lule älvdal [To Live on the River Side. Eight Researches about People and Resources in the Lule RiverValley]*. CEWE Publishers, Bjästa, pp. 57–77.

15 Bäcklund, D. (1988) *I industrisamhällets utkant. Småbrukets omvandling i Lappmarken [In the Borderland of Industrialization. The Transformation of Small Peasant Farming in the Lapland Area]*. Umeå Studies in Economic History no. 8, Umeå. Lundgren, N.-G. (1984) *Skog för export. Skogsarbete, teknik och försörjning i Lule älvdal [Timber for Export, Forest Work, Technology and Subsistence in the Lule RiverValley]*. Umeå Studies in Economic History no. 6, Umeå. Strand, H.-U. (1987) *Vattenkraften och befolkningsutvecklingen, sysselsättningen och ekonomin [Water Power and Population Development, Employment and the Economy]*. Vattenfall, Stockholm. See also Bäcklund (1996), note 4 above.

16 Tirén, L. (1937) *Skogshistoriska studier i trakten av Degerfors i Västerbotten [Forest Historical Studies in the Neighbourhood of Degerfors in Västerbotten]*. Meddelanden från statens skogsförsöksanstalt, Stockholm, Vol. 30(2). Stockholm. Tirén made a combined local study of forest history in the forest, mapping and timing the occurrence of forest fires, timber cutting and potash production with information about local economic history, e.g. numbers of population, arable land and cattle together with verbal reports and letters.

17 This concept has mainly been used for vertically integrated economic activities, e.g. the steel industry integrating with iron mines and engineering manufacturing; for example, in Sweden, an unsuccessful attempt with shipyards in the 1970s. Dahmén, E. (1970) *Entrepreneurial Activity and the Development of Swedish Industry, 1919–1939*. Irwin, Homewood. See also Dahmén, E. (1991) *Development Blocks and Industrial Transformation: the Dahménian Approach to Economic Development*. Almqvist & Wicksell, Stockholm.

18 At the Department of Economic History in Umeå, Erik Törnlund is doing a study of the closure of timber-floating in the Vindel river. Flottning, flottledsbyggnader och förändringar i vattenmiljön [Timber floating, floating construction work and changes in the water environment]. Umeå papers in Economic History no. 21, Umeå (1999). Also,

together with colleagues from the Department of Forest Vegetation Ecology, he has described the construction work over time in a small part of the river in order to analyse its interference with fish reproduction.

19 Agnoletti, M. (1998) *Sepherie e Foreste nel Trentino.* Museo degli Usi e Costumi, Trento.

Socioeconomical and Ecological Aspects of Coppice Woods History in the Lower Vosges (France) and the Black Forest (Germany)

9

R. Ostermann and A. Reif

Waldbau-Institut, Forstwiss. Fakultät der Universität Freiburg, D-79098 Freiburg, Germany

In the foothills of the Vosges (Alsace, France) recently abandoned coppice forests form a transitional belt between the high forests of the mountain range and of the adjacent vineyards and settlements. This study analyses the complexity of their history and natural preconditions. As the Black Forest is geographically related to the Vosges, the major features in techniques of coppicing and resulting stand structures of both sides are outlined. The main influences favouring coppicing in Alsace and Baden can be summarized as natural settings (climate and site conditions) on the one hand, and the historical background (history of settlement, history of forest utilization and forest policy) on the other hand.

1 Introduction

Today in Central Europe, coppicing is of secondary economic significance. Forestry in this region has to meet multiple demands that coppicing cannot accomplish any more. Nevertheless, large regions of recently abandoned coppice forests still exist, as in the foothills of the Vosges mountains of Alsace, which is included in the area under investigation. Today the Rhine is the political border between France and Germany. The Upper Rhine valley and the adjacent mountain ranges – which are the Vosges on the west side and the Black Forest on the east side – have a common origin and share similar landscape features and a similar land use pattern. But even though

(eds M. Agnoletti and S. Anderson)

this enclosed landscape has gone through a common history whose main trends are very similar, further consideration reveals systematic differences on both sides of the Rhine which imply significant differences in coppicing on both sides.

2 Area of investigation

The area of interest is shown in Fig. 9.1. Alsace and adjacent Baden are situated in Central Europe. Both regions are north–south oriented, parallel to the Rhine river. The landscape of Alsace can be subdivided into three units:

1. The *plain of Alsace* which is part of the Upper Rhine valley at an altitude of 250 m in the south and 150 m in the north.
2. The *foothills* that form a zone of transition from the plain to the mountains. They range between 250 to about 450 m a.s.l.
3. The *Vosges mountains* finally rising up to 1450 m in the south, but only 450 m in their northern part.

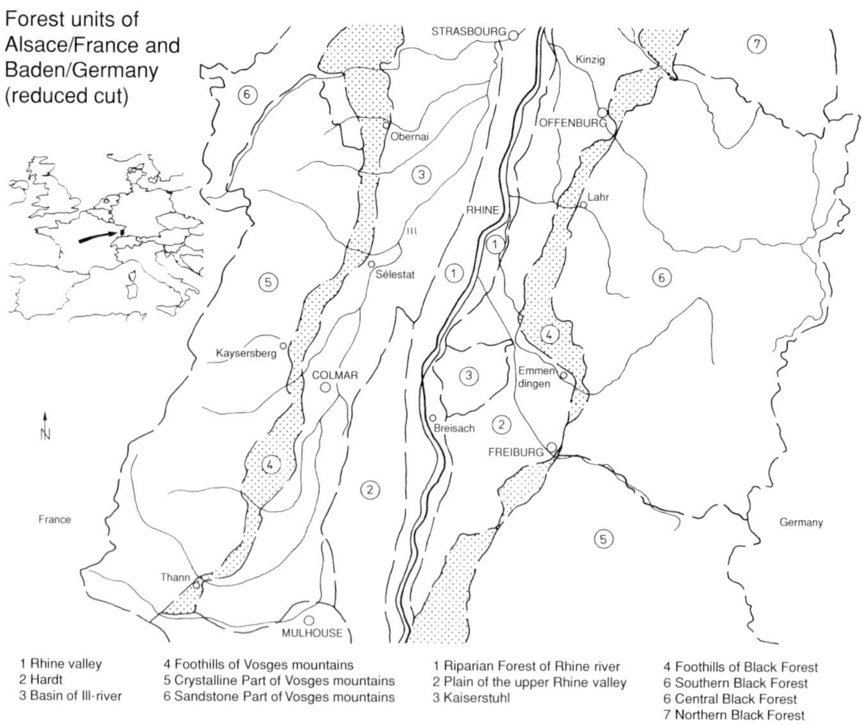

Fig. 9.1. The area of investigation.

The diverse landscape of Alsace shows a characteristic pattern of land use. In the plain of the Upper Rhine valley, agriculture with cereals and corn predominates. Deciduous forests occur here abundantly in the Ill Basin that extends parallel to the Rhine. Oak (*Quercus robur*) and hornbeam (*Carpinus betulus*) are the main canopy species in these coppice forests with standards. Throughout the foothills, wine is grown, often extending into the plain. Adjacent to the vineyards, at a higher altitude, on extreme relief sites and on northern slopes the forest of the Vosges mountains starts, reaching up to the ridges. These mountain forests formed by beech (*Fagus sylvatica*) and silver fir (*Abies alba*) are exploited as high forests. But along the eastern border of the Vosges mountains between the high-altitude forests and the vineyards occurs a small band of deciduous forests in a regular linear distribution, following the relief of the landscape. These forests are coppice forests.

On the eastern side of the Rhine the plain of Alsace is dominated by agricultural land use as well. Viniculture similar to that of the Alsace is common in the foothills of the Black Forest, but never extends into the plain. In some sections the foothills are dominated by agriculture or fruit growing. East of the foothills and geologically separated by several faultlines, high forest covers the slopes up to the ridges, composed of stands with pure beech, with mixed softwoods, hardwoods and conifers.

3 The present status of coppicing

In Alsace, the coppice forests are found throughout the foothills of central and southern Vosges. In the Black Forest they mainly occur in its central part. Moreover, north of the Kinzig valley, *Castanea sativa* coppice forests are found in the foothills of the Black Forest (Table 9.1). Throughout the northern parts of both mountain ranges, coppicing was never abundant as an acidic sandstone, poor in nutrients, dominates geologically. Therefore, they were thinly populated and land use was restricted to high forest exploitation.

One hundred years ago, in the foothills of the Vosges, coppice forests comprised more than 32,000 ha. In the equivalent foothills and inside the Black Forest they covered only about 20,000 ha. During the 20th century the coppice forests in Alsace were reduced by one-third and on the German side of the Rhine by two-thirds.

Other differences are related to forest use and the resulting stand structures. *Robinia pseudoacacia* coppice forests are encountered only in Alsace. *Corylus avellana* coppice forests and coppice forests mixed with *Betula pendula*, *Quercus petraea* and *Corylus avellana* typically occur only in the Black Forest region. In both regions *Castanea sativa* coppice forests and former tanbark coppice forests of *Quercus petraea* were frequent and still occur today.

Table 9.1. Comparison of coppicing in the investigated area.

	Alsace, France: eastern Vosges mountains and foothills	Baden, Germany: western Black Forest and foothills
Regional distribution of coppicing	Nearby to settlements in the foothills In a transitional belt between the lower situated vineyards and the high-altitude forests	In central Black Forest as part of an integrated land use system of individual farms In the foothills north of the Kinzig valley
Extension of coppice forests	1883: 32,739 ha* 1989: 18,524 ha‡	1850: 20,000 ha† 1989: 6,077 ha§
Actual types of coppice forests	Former tanbark coppice forests of *Quercus petraea* *Castanea sativa* coppice forests *Robinia pseudoacacia* coppice forests	Former tanbark coppice forests of *Quercus petraea* *Castanea sativa* coppice forests Former firewood coppice forests of *Betula pendula, Quercus petraea, Corylus avellana* *Corylus avellana* coppice forests

* Von Berg (1883).
† Abetz (1955).
‡ Ministère de l'Agriculture (1979–1989).
§ Mueller (1989).

4 Reasons for differences between the coppice forests in Alsace and Baden

Coppicing in Alsace and Baden can be related to physiographical attributes, including climatic and site conditions and to human influence during the past 2000 years. The human impact on the forests can be split up into three different characteristic historical components: the history of settlement, the history of forest utilization and the impact of forest policy.

4.1 Climate and site conditions

Alsace has a rather unusual climate compared to the rest of Central Europe. The Vosges mountains are the first brake for Atlantic clouds that drift eastward by dominating west winds. This means there is a high precipitation on the west

Table 9.2. Climate and sites.

	Alsace, France: eastern Vosges mountains and foothills	Baden, Germany: western Black Forest and foothills
Climate	Submediterranean to subcontinental: precipitation: 500–800 mm year^{-1}; temperature: 9.5–10.5°C	Subatlantic: precipitation: 900–2000 mm year^{-1}; temperature: 8–9°C
Site conditions	Few loessic deposits in the foothills, incorporated into thin residual loam on limestone Acidic granite and sandstone occur throughout the foothills Dominating steep slopes in the Vosges east of the main divide	Thick layer of loess in the foothills Acidic granite occurs in the foothills only north of the Kinzig valley Inside the Black Forest dominating plateaus

side of about 2000 mm year^{-1}, and low precipitation on the east side in Alsace at about 500 to a maximum of 800 mm year^{-1}. On the lee-side, 'foehn' often occurs preventing early and late frost in the foothills in autumn and spring. The average temperature per year ranges between 9.5 and 10.5°C. Summers are especially hot, and winters are mild. The climate can be generalized as submediterranean to subcontinental.

On the west side of the Black Forest, the weather side, ascending air condenses to rain clouds again, the precipitation rate rises according to the altitude up to 2000 mm year^{-1}. The average temperature is about 8–9°C, which characterizes this subatlantic climate (Table 9.2).

Soils in the foothills of the Vosges are derived from acidic granite, gneiss and sandstone, and locally from limestone. Aeolian loessic deposits are rare, and incorporated into a thin layer of loam in the topsoil. The foothills of the Black Forest are covered by a thick layer of loess, locally up to 10 m thick. North of the Kinzig valley, granitic bedrock is exposed.

4.2 Forest cover, structure and species composition

The submediterranean to subcontinental climate in the foothills of Alsace is naturally favourable to oak-dominated forests with *Quercus petraea*, and on calcarous soil *Quercus pubescens*. Beech (*Fagus sylvatica*) may have been mixed in locally. During Roman colonization, chestnut (*Castanea sativa*) was introduced. Subsequently it became an important component of the forests on acidic sites.

Quercus petraea and *Castanea sativa* are the main canopy species of the coppice forests. Because of different ecology and growth rates these tree species were grown in single stands. *Quercus petraea* still dominates today on southern slopes which are dry, rocky, of shallow soil and often degenerated by human processes due to intensive forest exploitation for centuries. *Castanea sativa* shows best growth on northern slopes with deeper soil with better water-holding capacity and of increased humidity. *Robinia pseudoacacia* coppice forests were raised irrespective of specific site conditions, but were almost exclusively limited to private forests.

In the foothills of the Black Forest area, climate is more subatlantic and humid. Forests naturally would be dominated by beech, mixed with fir above 600 m a.s.l. Inside the Black Forest *Fagus sylvatica* was only grown locally (>1000 m a.s.l.) successfully managed by coppicing. On other coppiced sites in the central parts of the Black Forest *Fagus* was eliminated and replaced by other species (*Quercus petraea, Corylus avellana, Betula pendula*).

Coppice forests were mainly formed by oak (*Quercus petraea*) and hazel (*Corylus avellana*). On the clearcut sites, birch (*Betula pendula*), as a pioneer species, became an important component of the canopy by natural succession. Coppice forests of these species occur east of the foothill zone. Because of extensive loess deposits, *Castanea sativa* coppice forests were raised only since the middle of the 19th century in the foothills north of the Kinzig valley on granitic bedrock.

4.3 Main features of human impact that favoured coppicing

History of settlement

During Roman colonization the foothills in Alsace were already a more favoured area for settlement compared to the eastern side of the Rhine valley. After the confused era of migrating nations since the 7th century, monasteries were established in the foothills. They were the original centres of steady colonization that followed. Towards the end of the Middle Ages Alsace had the highest density of population in Central Europe. Vine-growing landscapes typically could nourish more people than others (Dion, 1959). The cities and villages in the foothills were steadily extended and enlarged due to naturally rich soils combined with mild climate in a strategically important area. Since the Roman Age cereals were grown in the large plain of the Rhine valley, vines were grown in the foothills, and the mountains provided steady water and firewood to solve energy problems. As these settlements grew, sustainable forest management developed by coppicing close to the settlements where there existed a higher demand for a steady supply of firewood (short transportation distance was important!).

In contrast to the flourishing agriculture and viniculture in the foreland of the Vosges Mountains, its internal landscape remained hostile to colonization and farming. Only a few settlements were established in these unfriendly lands. Not until the 16th century did the silver and copper veins and the glass-blowing industry bring permanent settlers inside the mountains. The almost inexhaustible resources of beech and fir supplied these settlements sufficiently with firewood and timber. Exploitation for agricultural use, especially production of cereals, was too difficult because of the dominating steep slopes and the rugged character of the mountain range. Only the plateaus of the southern Vosges were exploited through the alpine pasture (Eggers, 1964).

The situation in the Black Forest was very different. After the turn of the millennium, a systematic opening of the mountains started after the foundation of monasteries and cities (Table 9.3). Plateaus and gentle slopes inside the Black Forest favoured agricultural land use and permanent settlements. Until today single farmhouses, typical 'Black Forest houses', are scattered throughout the valleys and plateaus with large private holdings of 40–200 ha forest, and more. In contrast to the southern Black Forest with higher altitude forests and pasturing, the farms of the central Black Forest region generated shifting cultivation called *Reutbergwirtschaft*. The shifting cultivation originally focused on agricultural production for self-sufficiency with cereals. The period of coppice forest was about 15 years, while forest pasturing was practised for up to 20–25 years. The brushwood of the coppice was only needed to fertilize the depleted soil after being burned to ash. Firewood was gathered

Table 9.3. History of settlement and agricultural land use.

	Alsace, France: eastern Vosges mountains and foothills	Baden, Germany: western Black Forest and foothills
History of settlement	Since 7th century consolidation of settlements around Vosges mountains, no permanent settlements inside the mountains	At the turn of the millennium systematic opening of the mountains from outside and with outposts inside
Agricultural land use in the mountains	Alpine pasture, only few permanent settlements; locally *Brandfeldbau* without a period of coppicing	In central Black Forest shifting cultivation practised by the farms, *Reutbergwirtschaft*, for self-supply with basic food, including a period of coppicing

and removed from different stands. Thereafter, burning followed a period of 2–3 years of agricultural land use; rye was particularly popular and, since the 18th century, potatoes were grown until the soil was depleted in nutrients and the sprouting stumps again reconquered the terrain.

Inside the Alsatian Vosges, shifting cultivation with a short period of coppice forest was not common. Only in some areas such as the upper Bruche valley and the upper Val de Villé was *Brandfeldbau* or *essartage* practised. In contrast to the *Reutbergwirtschaft* of the Black Forest, *Brandfeldbau* was practised on the public property of a village. Portions of land were distributed to the villagers for a period of agricultural use. After the depletion of the soil, fallow land followed for 10–20 years. Sometimes the land was pastured, if not, shrubs such as broom (*Sarothamnus scoparius*) and heather (*Calluna vulgaris*) reconquered the areas until they were burned again to fertilize the next crop. A short period of coppicing was not common (Krzymowski, 1914).

History of forest use

Various commercial uses for small coppice wood were common in Alsace. Originally, and continuously until after World War II, coppicing had to satisfy the need of firewood, especially close to the densely populated foothills with its cities and villages. In the Black Forest during the 19th century, when shifting cultivation was abandoned after radical change of socioeconomic structures, firewood coppice became common (Table 9.4). In the foothills, they may have existed locally.

As initially mentioned, coppice forests in the Alsace were grown close to vineyards. In fact there is a long tradition and significant relation of coppicing and stickwood production for viniculture that is rooted in Roman colonization. From the time the Romans introduced the technique of viniculture in Alsace until the 1950s it was a tradition to fix every single vine with a stick, using 6000–7000 pieces ha^{-1}. Since the 13th century, instead of the regular cubic metre, the foresters used an extra measure unit for sticks which was $2.6 \times 1 \times 1$ m (Kahl, 1894). Today's mechanized methods commonly use wire and concrete poles.

Originally oak sticks were used, as *Quercus petraea* is indigenous and its wood is weatherproofed. Since the 17th century, chestnut, *Castanea sativa*, was cultivated as coppice forest and replaced the oak stickwood. *Castanea sativa* was introduced during Roman colonization but probably it was more important in Alsace as a source of food than for stickwood, although the Romans already knew about coppicing of chestnut. When chestnut coppicing became important in the 18th century the potato had already been introduced to Central Europe and replaced the chestnut as a basic food (Livet, 1986). In the 19th century another tree species became popular, *Robinia pseudoacacia*; it was important as stickwood because of its weatherproofed quality.

Table 9.4. Coppice forest products.

	Alsace, France: eastern Vosges mountains and foothills	Baden, Germany: western Black Forest and foothills
Firewood	Originally and continuously until after World War II	Beginning in the 19th century when shifting cultivation was abandoned Replaced by firewood coppicing
Stickwood production for vineyards	All along the foothills until the 1950s Originally oakwood; from 18th century chestnut; from 19th century *Robinia*	Only chestnut in the foothills north of Offenburg from 1870 to 1950; only of local significance
Tanbark coppicing	From 1850 to 1900 commercial tanbark coppicing promoted by governmental support	From 1850 to 1900 commercial tanbark coppicing promoted by governmental support on abandoned sites of shifting cultivation

Along the foothills of the Black Forest vine growing was also common, but stickwood production by coppicing was not adopted. Often fir poles were used, especially the young trees that grew up in the understorey of the forest which were more weatherproofed because of their small tree rings (Mueller, 1938). Many of the poles produced in Alsace were sold to the vineyards in Baden. Beginning at the end of last century, chestnut coppicing was introduced to the loess-free part of the foothills north of the Kinzig valley, a practice that appears to have been copied from Alsace (Soelch, 1950).

Oak bark of *Quercus petraea* was the main tanning agent used until the late 19th century. About 1850, there began a period of rapid increase in the demand for tanbark. Over the next 50 years high prices made tanbark coppicing a very valuable form of land use in Alsace, in the central Black Forest and in other regions of Central Europe. The high quality of *Quercus petraea* as firewood also boosted the value of tanbark coppicing. After the turn of the century foreign oak bark became increasingly important, eventually being replaced by chemical substitutes when tanbark coppicing was abandoned. This development was the same on both sides of the Rhine.

Forest policy

France has different ecological settings for forest growth compared to Central Europe: the Atlantic climate in flat northwestern France favours natural

Table 9.5. Forest policy.

	Alsace, France: eastern Vosges mountains and foothills	Baden, Germany: western Black Forest and foothills
Forest policy	Forest law of 1669 prescribed coppicing Changing influences of German and French forestry depending on the political hegemony	Different forest law in every sovereignty After 1800, scientific forestry promoted forest management in high stands
Beginning of transformation in high stands and conifer stands	Up to World War I, then again after the 1960s	Steadily since the decline of tanbark coppicing at the end of the 19th century

deciduous forests of beech (*Fagus sylvatica*), oak (*Quercus petraea* and *Q. robur*) and hornbeam (*Carpinus betulus*). Since the early Middle Ages coppicing was spread throughout these regions and was often prescribed in forest laws (Table 9.5). It was Minister Colbert who invented the famous Forest Ordinance of 1669 during the reign of Louis XIV. Following this, coppicing was imperative in all public forests to guarantee sustained-yield management of all French forests, including the Alsace, which was annexed to France after the 30 years war in 1648. Because of this famous law, coppicing became the main kind of forest management in most deciduous forests until the middle of this century in all France, in addition to the afforestation activities that began in the 19th century with conifers in some regions of the country (Devèze, 1966).

While the administration of France is still centrally organized, the German territory was split up into sovereignties, each of them having its own forest laws designed for local needs and local natural settings, as in the rich coniferous forests of the highlands of Central Germany (Table 9.5). The industrialization that began in the early 19th century changed the demand for forest products: trees were cut for construction purposes and pulpwood instead of for firewood, especially as charcoal became a source of fuel. The forest administration quickly adopted intensive forest management followed by the cultivation of coniferous forests (*Pinus sylvestris* and *Picea abies*).

This change in Germany at the beginning of the 20th century in the demand for forest products and in the decline of the tanbark industry led to an increased abandonment of coppicing. As a result, coppice forests all over the Black Forest have been gradually transformed. During the 1960s–1970s the transformation was almost completed when the forest owners were given governmental financial support (Schuelli, 1967).

Alsace suffered from the changing influences of German and French forestry (German territory until 1648, French: 1648–1870, German: 1870–1918, French: 1918–1939, German: 1939–1945, French since 1945). Both nations have held widely differing concepts of forest management since the 17th century. While France maintained coppicing almost until the 1960s, German forest administration abolished coppicing successively owing to the collapse in profitability. In the early 1960s in Alsace, there began a very late transformation of coppice forests. Thereafter began a movement of 'enrésinement', which in France since the beginning of the 20th century has meant afforestation with conifers as well as the replacement of deciduous trees with conifers (transformation). Recently, coppice forests of oak (*Quercus petraea* and *Q. pubescens*) have become important for nature conservation as they shelter distinct plant communities of high floristic value (a submediterreanean exclave in Central Europe).

Conclusion

In Alsace since the early Middle Ages, coppicing was improved and cultivated as this method of forest management had to accomplish multiple demands in forest products – first of all firewood and stickwood. For these reasons coppice forests were grown adjacent to the vineyards and close to the settlements. Beginning in the second decade of the 19th century, tanbark coppicing boomed followed by a rapid decline. Today's new methods of vineyard cultivation make stickwood useless and firewood is only of local interest. For almost 50 years coppicing has been abandoned as its economic use is limited. The transfomation of the forests is a very slow process. Consequently it can be seen that coppicing in the foothills of Alsace has a long tradition. Today's coppice forests still reflect their former intensive exploitation.

In contrast with Alsace, Baden has no similar tradition in coppicing. The shifting cultivation of the central Black Forest area focused on agricultural production. When this land use was first abandoned, the Black Forest experienced a short period of coppicing (firewood and tanbark, and local stickwood production), but after the turn of the 20th century, coppicing was successively abolished by the forest administration. This change in forest land use is still evident today.

The main reasons for the different development on both sides of the Rhine are found in natural settings and in the historical background. Both must be considered to understand coppicing. As this study was carried out as a contribution to the understanding of forests in the cultural landscape it has focused on the interactions between forests and people. Together, these two factors have created a unique cultural landscape in the Upper Rhine valley and its adjacent mountains.

References

Abetz, K. (1955) *Baeuerliche Waldwirtschaft.* Parey, Hamburg and Berlin.

Barth, M. (1958) *Der Rebbau des Elsasses und die Absatzgebiete seiner Weine. Ein geschichtlicher Durchblick.* Roux, Strasbourg and Paris.

Devèze, M. (1966) Forêts françaises et forêts allemands: Étude historique comparée. *Revue Historique* 479, 347–380 and 480, 47–68.

Dion, R. (1959) *Histoire de la vigne et du vin en France des origines au XIXe siècle.* Paris.

Eggers, H. (1964) *Schwarzwald und Vogesen. Ein vergleichender Überblick.* Westermann, Braunschweig.

Kahl, A. (1894) *Forstgeschichtliche Skizzen aus den Staats- und Gemeindewaldungen von Rappoltsweiler und Reichenweier.* Beiträge zur Landes- und Volkskunde 19. Heitz & Muendel, Strassburg.

Krzymowski, R. (1914) *Die landwirtschaftlichen Wirtschaftssysteme Elsass-Lothringens.* Gebwiller, Elsaß.

Livet, G. (1986) La campagne alsacienne au carrefour des influences (XVIIe et XVIIIe siècle). In: Livet, G. (ed.) *L'Europe, l'Alsace et la France: Problèmes intérieures et relations internationales à l'époque moderne.* Editions d'Alsace, Colmar.

Ministère de l'Agriculture (1979–1989) *Résultats de l'Inventaire Forestier, Département du Bas-Rhin,* Vol. 1. *Département du Haut-Rhin,* Vol. 1. Paris.

Mueller, K. (1938) *Geschichte des Badischen Weinbaus.* Schauenburg Verlag, Baden.

Mueller, G. (1989) Niederwaldreste und Weidfeldsukzessionswald im Mittleren Schwarzwald. Zustand und Behandlungsempfehlungen. Unpublished manuscript, Forstdirektion Freiburg.

Schuelli, L. (1967) Aufbau und Umwandlungen in den Bauernwaldungen des mittleren Schwarzwaldes von 1850–1960. *Schriftenreihe der Landesforstverwaltung Baden-Wuerttemberg* Vol. 24. Stuttgart.

Soelch, G. (1951) *Der Kastanien-Niederwald.* Unpublished manuscript, Office of the Forest District, Bad Peterstal.

Von Berg, K. (1883) *Die forstwirtschaftlichen Verhaeltnisse in Elsass-Lothringen.* Strassburg.

The Forest History of Boreal Sweden: a Multidisciplinary Approach

10

L. Östlund and O. Zackrisson

Department of Vegetation Ecology, Swedish University of Agricultural Sciences, S-90183, Umeå, Sweden

The forest history of boreal Sweden can, with special regard to the human influence on the forest, be divided into three distinctive phases: an early period dominated by agricultural use of the forest ecosystem, an intermediate period with a diverse forest use including agricultural and preindustrial activities, and finally an industrial phase. The shift from the first phase to the third occurred within the 18th and 19th centuries. During this period the forest of northern Sweden has been transformed, basically, from a forest predominantly generated by natural processes towards a forest strongly influenced by human activities. Agricultural, preindustrial forest and early industrial forest utilization was generally aimed at specific resources from the forest, i.e. leaf fodder, pine stumps or very large pines. In the 20th century, forest utilization and forest exploitation have been replaced by forest management. This has increased the overall growth rate and standing volume of the north Swedish forest, but has also had an ecological price. During the period of intensive silviculture in the 20th century, biologically important structures have been consistently removed from the forest and biological diversity has been reduced. Forest historical studies will allow us to understand long-term changes in ecologically crucial structures and components of the forest landscape. It will also help us to understand changes in forest ecosystem dynamics and functions. A large number of primary historical sources as well as biological archives can be used to answer such questions. However, the use of historical sources requires a good understanding of source criticism as well as knowledge of forest ecology and forest management. We believe that such research can not be completed successfully without integrating methods and theory from different branches of science.

1 Introduction

Sweden lies on the periphery of Europe. There is a characteristic climatic gradient northwards which is mirrored by decreasing population density. This is important to keep in mind as a background for understanding the forest history of this part of Europe. The northern, boreal part of Sweden is a land of few people but large forests. Most of this part of Scandinavia is covered by forest, with approximately 60% of the land area forested. Only a few per cent is agricultural land; the rest is wetlands with or without sparse forest cover and mountains above the timberline (Östlund, 1993).

The dominating tree species include Scots pine (*Pinus sylvestris*) and Norway spruce (*Picea abies*), the only two important native conifer species. Together they comprise almost 90% of the standing timber volume. Apart from these conifers there are also several species of deciduous trees, primarily birches, but also aspens and a few others, but they only account for 10% of the standing volume.

The forest history of this part of Sweden is rather dramatic and the use of the forest resources has been a very important component in Sweden's transformation from an agricultural nation in the 19th century to an industrialized country in the 20th century. The major influence on the boreal forest prior to the 19th century was connected with various agricultural practices (Zackrisson, 1976; Ericsson and Östlund, 1998). The most important activities were grazing by domestic animals in the forest during the summer and the collection of fodder for the animals for the winter. To improve grazing the natural fire frequency was altered and we now believe that fires occurred more frequently in some forest types than they would have under more natural conditions. This probably affected long-term regeneration patterns and the relative abundance of Scots pine versus Norway spruce, since the pine species is rather fire-tolerant compared to the more fire-sensitive spruce.

Wildfires and set fires were common in boreal Sweden until the early 20th century. Approximately 1% of the forest land burnt annually until fire suppression became effective in the early 20th century (Zackrisson, 1977). Many species of plants and animals are adapted to fire in these ecosystems and the long period of fire suppression has resulted in serious biodiversity problems.

2 Preindustrial forest exploitation: tar and potash

In the 19th century northern Sweden came into contact with the international market for the first time. Two different preindustrial forest uses became important. Potash came to be a rather important export item from northern Sweden during approximately 50 years in the 19th century (Tirén, 1937; Östlund *et al.*, 1998). Potash was produced by burning wood of deciduous trees, primarily birches and aspens, in large piles. The ash was then collected and one chemical, potassium carbonate, was extracted from the ash by a relatively simple process. There is an approximate ratio between potash and wood of 1:1000. The potash could easily be transported and sold at a high price. Most of it was exported abroad to glass industries in the industrialized parts of Europe. The production of potash influenced the forests in an important way. The preferred raw materials were old, dead or half-rotten deciduous trees. Today such trees carry an important biological legacy and many rare species of fungi, insects and birds are highly dependent on such trees.

The other important product was wood tar (Borgegård, 1973). The tar was produced in special tar kilns. This wood tar was transported in barrels and again primarily exported from Sweden. Before 1809, when Sweden and Finland were one country, the major part of the tar production took place in the northern Finnish provinces. In these areas pines were deliberately scarred to produce tar substances and then felled after a few years. When Sweden and Finland separated, production also began in the northern part of Sweden. In this case the raw materials were dead pine trees and snags killed by the recurrent fires.

3 Logging and forest management in the 19th and 20th centuries

Even taking into account what we have said earlier about agricultural use of the forest and different preindustrial activities, the major part of the forest of northern Sweden was relatively little influenced by man until the 1850s (Östlund *et al.*, 1997). At that time the first major landscape-encompassing human impact on the forest began, and this part of Europe was very rapidly brought into the industrial revolution. A so called 'timber frontier' moved across the landscape, driven by demand from the industrially developed countries in Western Europe (Björklund, 1984). The need for sawn and square-cut timber was great and these items were used for shipbuilding, factories, workers' housing, sleepers for the railroads and many other things. Around 1870 Sweden was the world's largest exporter of sawn products and all of the timber came from the northern provinces.

The target for the exploitation was the old-growth pine-dominated forest. Trees of 300–500 years were common in these forests (Linder and Östlund, 1998). This forest type was clearly shaped by the recurrent low-intensity forest fires. The largest pines survived the fires and even benefited from the nutrient release after the fires. The exploitation coincided with an agricultural expansion in northern Sweden. The new farmers did not succeed so well in farming but they were extremely important as loggers in the forests instead.

To extract the timber the dense network of rivers in northern Sweden was used and by the beginning of the 20th century there were many more miles of floatways in northern Sweden than there were roads. The use of the rivers and streams for the floating has changed these in a rather dramatic manner. Rivers have been cleared and different constructions, such as timbered box booms and stone piers, are still remnants of this period (Hellstrand, 1980; Östlund, 1995).

After the turn of the 20th century, the forest exploitation changed into a more sophisticated forestry. In 1903 the first modern forestry law was passed in Sweden (Stjernqvist, 1973). This law stressed sustainable yield and has been quite successful during the 20th century. Both the total timber volume has increased and the logging has increased during almost 100 years.

It is difficult to summarize development during the 20th century in simple terms but the extensive planting of seedlings of the preferred species (primarily Scots pine) and the extensive removal of unwanted species (primarly deciduous trees), either by thinning or by other means such as large-scale use of herbicides, has been very important. Also we have experienced a very significant increase in mechanization. Forty years ago men felled the trees with either hand-saws or the earliest chain-saws and horses were used for getting the timber out of the forest. Today all logging operations are done by machines. A dense network of logging roads has also been built. The focus on timber production during almost a century has had a high ecological price. Today we have many red-listed species of animals and plants. The present movement is towards environmentally safe forestry but there are doubts as to whether it is too late to save some species and ecologically valuable forests. A very small portion of the Swedish forest is currently protected and set aside from forest exploitation.

4 Sources for the study of forest history in Sweden

Some of the questions a forest historian might want to ask concern the natural development of forest vegetation, forest exploitation, to discern human-induced and natural changes of forest ecosystems and to analyse ecological effects of human changes on forest ecosystems. There are a large number of primary historical sources as well as biological archives which can be used to answer such questions. However, the use of historical sources requires a good understanding of source criticism as well as knowledge of forest ecology and forest management. The former is especially true for the ecologist and the latter is just as important for the historian and for the economic historian.

One of the major problems which has to be dealt with when the forest historian uses historical records concerns the interpretation of information. Measurements of timber and various forest products have not been consistent throughout history, but have varied considerably over time and between different regions. This makes it very important that the forest historian clearly understands the background of his/her sources before analysing quantitative information.

The interpretation problems may seem as rather small and only causing problems for those studying these things in detail. Often, however, it is important to estimate local and regional impact of different human activities on forest ecosystems (Wieslander, 1936; Arpi, 1951). Even a small error in the primary calculations can result in a complete misinterpretation of the volumes of wood used for charcoal production, potash burning or saw-timber logging and the corresponding impact on the forest ecosystems. Our present understanding of human impact on the preindustrial forest ecosystems is based in many cases on insufficient and incomplete analyses of the use of forest resources in the past.

Another important problem concerning historical records is gaps in the records. Peterken (1979) has pointed out that scientists, skilled in different scientific methods and used to carefully controlled experiments or investigations which can be statistically analysed, are not used to gaps or inconsistencies in historical records. There may be several reasons for these gaps. Sources might never have been produced, or they were destroyed deliberately or accidentally (Cipolla, 1991). In the case of records deriving from logging companies and the metal industries of northern Sweden, the archives are generally relatively intact and complete compared with other countries. Smaller or larger gaps in different records do occur but often this can be dealt with because similar information can be collected from other kinds of records, thus making it possible to fill in gaps in one series of records.

Borgegård (1973) discusses the problem of gaps in the records in his thorough investigation of tar production in the county of Västerbotten during the 19th century. One of his objectives was to compare the purchase of tar from farmers by different tar companies in the region. While it was possible to estimate the deliveries of tar to one specific company for an extended period of time, it proved difficult to compare different companies for any length of time. His conclusion is that gaps in the source material often restrict the possibilities to study longer time periods.

This particular case also highlights a paradoxical problem in studies of forest history of the 18th and 19th centuries. Rapid economic development combined with radical institutional changes during the 19th century generated a number of new companies and individuals dealing in timber and other forest products. Before that, monopoly conditions were more common, i.e. only one company had control over forest exploitation in one area and excluded all competition. Therefore, although the source material generally gets better in quality and in quantity over time, it is in some cases easier to obtain a more complete picture of forest exploitation for an earlier period, when often there were only one or few actors involved. This is especially true when it comes to timber exploitation along the major rivers of northern Sweden in the decades before and after 1850 (Östlund, 1993).

An important exception to the rule that the larger lumber company archives are usually without major gaps is the case of one of the largest water-driven sawmills of northern Sweden in the 19th century, the sawmill at Baggböle in the county of Västerbotten. The owner of the sawmill, James Dickson & Co., was involved in many legally doubtful activities from the 1840s to the 1870s. The company was accused of promoting theft of timber from public forests by paying farmers to cut timber illegally (Gaunitz, 1980; Bunte *et al.*, 1982). The legal proceedings almost inevitably acquitted the representatives from the charges in the many trials, but the company probably deliberately destroyed large portions of the critical timber ledgers in order to avoid further legal actions (Balgård, 1980). Much of the research of this particular event therefore has to be based on secondary sources or records from other archives.

The study of the industrial use of the forest is most often based on records produced by a logging company or by the authorities. This gives an unwanted 'one-sidedness' to the rendering of the exploitation as the process is only regarded from one perspective. The opinions of the farmer selling the timber or the logger cutting it or the floater moving the logs are seldom recorded. However, there are records in which it is possible to trace the 'other side' of the forest exploitation process. Such relatively rare but nevertheless important set of records are peasant diaries, farmer diaries and other similar sources which have been presented earlier. Court proceedings and court investigations can be studied to investigate the opinions and views of the farmers and workers through their replies to court charges and their recorded participation in trials as defendants or witnesses (Kardell, 1992). The value in using these types of records lies not only in the wider perspective they provide, but also in the possibility to check the validity and bias in other records. It is, furthermore, obvious that the everyday domestic use of the forest resources is consequently underestimated due to the scarce source material and the difficulties in interpreting the available sources (Kardell, 1991).

Debates concerning the available forest resources and sustainable yield from forests have almost invariably been an integral part of Swedish history. The reasons for this are both the importance of forest resources for the national economy as well as the forests continuously being a dominant feature of the Swedish landscape. During the 16th century, when a strong centralized government arose, one important issue was the ownership of the vast northern forests of the country. A delineation process was started which resulted in transfer of forest land from the state to private farmers (Almquist, 1928a). In the 17th century, several Royal forestry commissions tried to sort out problems of property rights and sustainable yield in the common forests of southern Sweden (Almquist, 1928b). During the 18th and 19th centuries steadily increasing industrial utilization of the forests again brought conflicts over the ownership of land, when sawmill owners bought forest land from independent farmers of northern Sweden (Holmgren, 1959; Fagerberg, 1973; Sörlin, 1988).

Many of the reports from the different commissions and from officials are biased, reflecting the opinion that the forest had to be protected from human exploitation and predicting a future without enough forest resources for the inhabitants (Mattsson and Stridsberg, 1981). Therefore, official reports from this time have to be analysed with great care and with a thorough examination of the basis for the work of the commissions. Much of the 20th century debate about the historical state of the forest in Sweden has unfortunately been based on an uncritical acceptance of clearly biased statements from the 19th century about forest resources (Hamilton, 1983; Remröd, 1990). This question has, however, been thoroughly discussed among others by Nordström (1959) and Linder and Östlund (1992).

A central issue of forest history is to describe the state of a forest at different times. This can be achieved by using a variety of historical sources, including for example forest censuses, forest surveys and other survey records. The ideal situation would be to compare forest surveys, which were carried out identically at different times throughout history. This is not usually the case, however, especially when a time perspective of more than 100 years is desired. It is not necessarily so that the earlier records are less detailed or accurate, but the problems most often relate to the different objectives used when the records were set up (Watkins, 1984). Other sources may pose more difficult problems and the forest historian also may have to analyse sources which are quite different. This reality has to be accepted and discussed, and not regarded as an insuperable obstacle. Forest historical sources, like historical sources in general, may be seen as circumstantial evidence in a trial and not as reliable eyewitness accounts (Peterken, 1979). The forest historian nevertheless has to build a strong case on the information he has available. He also has to fight the 'viewpoint' imposed on him by the sources (Cipolla, 1991).

5 Combining biological archives and historical records

We believe that a broader and better understanding of the Swedish boreal forest history must be based on knowledge from several different disciplines. This does not only imply that knowledge from several disciplines should be used, but also that it is necessary to combine methods and theories from different disciplines. On one hand, analysis of historical records can provide a detailed insight into the documented human use of the forest. On the other hand, the use of biological archives (analysis of pollen, tree rings, etc.) can provide information on subjects when the documentary records are silent. The natural history of forest ecosystems and the human-induced changes in these, together form a complicated and interwoven pattern which can not easily be separated (Hörnberg *et al.*, 1999). We believe therefore that it is necessary to accept that the interpretation of forest history will benefit if a multidisciplinary approach is used.

Our present knowledge of the forest history of boreal Sweden still has major gaps despite a long tradition of research by scholars of different disciplines. In our opinion some areas of research are especially important to pursue in the future:

1. Preindustrial forest exploitation including small- and large-scale use of specific forest resources such as charcoal wood, potash wood and pine stumps for tar production are of great interest. We believe that earlier conclusions about the impact on the forest ecosystems are sometimes drawn from insufficient material.

2. Ecological studies focusing on the transformation of the boreal forest landscape during the last 200 years is another important example. Despite the radical transformation of the forest and its ecological consequences, we still know very little about the pre-exploitation state of the forest and the exploitation process.

3. Detailed studies of the long-term changes of the north Swedish forest including natural and human-induced changes are of special interest. The challenge in this case is to connect biological archives with historical sources, i.e. to connect the long-term perspective with a shorter perspective.

The first field of research represents a revision and evolution of research in the past, while the latter two fields of research could produce knowledge of particular importance at the present time. Forest managers and nature conservationists in Sweden definitely lack knowledge about long-term stability and change in forest ecosystems. While the objectives for these studies may vary, we believe that such research can not be completed successfully without integrating methods and theory from different branches of science.

Acknowledgements

The Bank of Sweden Tercentenary Foundation and the Swedish Council for Forestry and Agricultural Research provided funds for this work.

References

Almquist, J.E. (1928a) Det norrländska avvittringsverket [Delineation in northern Sweden]. In: Anon. (ed.) *Svenska Lantmäteriet 1628–1928*. Sällskapet för Utgivande av Lantmäteriets Historia, Stockholm, pp. 365–494.

Almquist, J.E. (1928b) Skogskommissionerna under det karolinska enväldet [The forest commissions during the Caroline autocracy]. In Anon. (ed.) *Svenska Lantmäteriet 1628–1928*. Sällskapet för Utgivande av Lantmäteriets Historia, Stockholm.

Arpi, G. (1951) *Den svenska järnhanteringens träkolsförsörjning 1830–1950* [*Charcoal supply of the iron industry in Sweden 1830–1950*]. Geografiska Institutionen, Uppsala University, Uppsala.

Balgård, G. (1980) Ödemarken, som bevittnat dåden är stum [*The wilderness, witnessing the crimes, remains silent*]. *Västerbotten* 1, 15–40.

Björklund, J. (1984) From the Gulf of Bothnia to the White Sea – Swedish direct investments in the sawmill industry of Tsarist Russia. *Scandinavian Economic History Review* 32, 18–41.

Borgegård, L.E. (1973) *Tjärhanteringen i Västerbottens län under 1800 – talets senare hälft* [*Tar trade in Västerbotten county during the late 19th century*]. Kungl Skytteanska Samfundet, Umeå.

Bunte, R., Gaunitz, S. and Borgegård, L.E. (1982) *Vindeln – En norrländsk kommuns ekonomiska utveckling 1800–1980* [*Vindeln – economic growth in a rural district of northern Sweden 1800–1980*]. Selector, Lund.

Cipolla, C.M. (1991) *Between Two Cultures. An Introduction to Economic History*. Norton, New York.

Ericsson, S. and Östlund, L. (1998) A forest of grazing and logging. The history of a boreal forest in northern Sweden. *New Forest* (in press).

Fagerberg, B. (1973) The transfer of peasant forest to sawmill companies in northern Sweden. *Scandinavian Economic History Review* 11, 164–191.

Gaunitz, S. (1980) Baggböleriet. *Västerbotten* 1, 2–14.

Hamilton, H. (1983) *100 år i skogen* [*A hundred years in the forest*]. Sveriges Skogsvårdsförbund, Stockholm.

Hellstrand, G. (1980) *Flottningen i Dalälven* [*Floating on Dalälven River*]. Stiftelsen Dalarnas Museum, Falun.

Holmgren, A. (1959) Skogarna och deras vård i övre Norrland intill år 1930 [The forests and their care in the northernmost part of Sweden to 1930]. In: Arpi, G. (ed.) *Sveriges skogar under 100 år*. Domänstyrelsen, Stockholm, pp. 375–443.

Hörnberg, G., Östlund, L. and Zackrisson, O. (1999) The genesis of two *Picea–Cladina* forests in northern Sweden. *Journal of Ecology* 87, 800–814.

Kardell, L. (1991) En skogshistorisk skiss [A sketch of forest history]. In: Pettersson, T.J.E. (ed.) *Lima och Transtrand. Ur två socknars historia*. Malungs Kommun Malung, pp. 103–182.

Kardell, L. (1992) Skogshistoriska notiser ur Gotlands dombok år 1836 [Notices of forest history from Gotland county's judgement book of 1836]. *Skogshistorisk Tidskrift* 1, 50–67.

Linder, P. and Östlund, L. (1992) Förändringar i norra Sveriges skogar 1870–1991 [Changes in the boreal forests of Sweden, 1870–1991]. *Svensk Botanisk Tidsskrift* 86, 199–215.

Linder, P. and Östlund, L. (1998) Structural changes in three north Swedish forests 1881–1997. *Biological Conservation* 85, 9–19.

Mattsson, L. and Stridsberg, E. (1981) *Skogens roll i svensk markanvändning – En utvecklingsstudie* [*The role of the forest in the landuse of Sweden – a development study*,]. Institutionen för Skogsekonomi, Sveriges Lantbruks Universitet, Umeå.

Nordström, L. (1959) Skogsskötselmetoder och skogslagstiftning [Methods of silviculture and legislation relating to forests]. In: Arpi, G. (ed.) *Sveriges skogar under 100 år*. Domänverket Svenska Skogsvärds Foreningen, Stockholm, pp. 241–262.

Östlund, L. (1993) *Exploitation and structural changes in the north Swedish boreal forest 1800–1992*. Thesis, Swedish University of Agricultural Science, Department of Forest Vegetation Ecology, Umeå.

Östlund, L. (1995) Logging the virgin forest: northern Sweden in the early-nineteenth century. *Forest and Conservation History* 39, 160–171.

Östlund, L., Zackrisson, O. and Axelsson, A.L. (1997) The history and transformation of a Scandinavian boreal forest landscape since the nineteenth century. *Canadian Journal of Forest Research* 27, 1198–1206.

Östlund, L., Zackrisson, O. and Strotz, H. (1998) Potash production in northern Sweden – history and ecological effects of a pre-industrial forest exploitation. *Environment and History* 4, 345–358.

Peterken, G.F. (1979) The use of records in woodland ecology. *Archives* 14, 81–87.

Remröd, J. (1990) *Möjligheternas skog* [*The forest of possibilities*]. *Skogsindustrierna*. Media Express, Stockholm.

Sörlin, S. (1988) *Framtidslandet – Debatten om Norrland och naturresurserna under det industriella genombrottet* [*The land of the future – public discussions on northern Sweden and natural resources during the industrial revolution*]. Carlsson, Malmö.

Stjernquist, P. (1973) *Laws in the Forest*. Acta Regiae Societatis Humaniorum Literatum Lundensis, Lund.

Tirén, L. (1937) Skogshistoriska studier i trakten av Degerfors i Västerbotten [Studies of forest history near Degerfors in Västerbotten]. *Meddelanden från Statens Skogsförsöksanstalt* 30, 67–322.

Watkins, C. (1984) The use of Forestry Commission censuses for the study of woodland change. *Journal of Historical Geography* 10, 396–406.

Wieslander, G. (1936) Skogsbristen i Sverige under 1600- och 1700-talen [Forest scarcity in Sweden during the 17th and 18th centuries]. *Sveriges Skogsvårdsförbunds Tidskrift* 34, 593–663.

Zackrisson, O. (1976) Vegetation dynamics and land use in the lower reaches of the river Umeälven. *Early Norrland* 9, 7–74.

Zackrisson, O. (1977) Influence of forest fires on the north Swedish boreal forest. *Oikos* 29, 22–32.

Preindustrial Forests in Central Europe as Objects of Historical–Geographical Research

11

W. Schenk

Geographisches Institut der Universität Tübingen, Hölderlinstraße 12, 72074 Tübingen, Germany

Forest changes in time and space, whether they occur naturally or anthropogenically, influence in their turn the balance of nature as well as the possibilities and strategies of value setting by man. The resulting research problems are complex and lie between natural science and the humanities. Starting from their own specific approach, a multitude of disciplines deal in varying intensity with the forests as a historical phenomenon. In this chapter an attempt is made in the form of a research report to put order into the cornucopia of historical–geographical literature in Germany, understood as the production of a subdiscipline of geography and adjacent disciplines, on the development of forests in preindustrial time (i.e. 16th century until the 1850s) and, based on this, to outline specific contributions and possibilities of a historical–geographical approach dealing with forests.

1 The forest as research object for historical geography?

The large amount of forest in the cultural landscapes of Central Europe (Troll, 1962) makes it *per se* an important phenomenon for geographical research (Rubner, 1928). However, a general survey of geographical literature of the past 30 years, since the publication of the only textbook so far on forest and woods' geography by Windhorst (1978a,b) shows only a small number of relevant works compared with the total production of this discipline. If they are regarded explicitly as works from this subdiscipline then they stem mostly from Windhorst himself or from his immediate surroundings (Klohn and Windhorst, 1984; Klohn, 1986). In spite of Windhorst's actualistic and global conception, forest and woods' geography has hardly developed as a subdiscipline since the 1970s. But this does not mean that geography has turned its back on the forest as an object for research. Rather, it is an integrated part of numerous geographical works and is considered under the heading of the geographical subdiscipline of which this researcher feels himself a part.

As for geography in general, the same is true for the position of the forest in generic cultural landscape research (Jäger, 1987). Explicit research on the forests in their historical–geographical condition (Jäger, 1968) has remained rare in the past three decades in comparison with the total production of the discipline.* However, in many works on regional cultural landscape development (Röll, 1966; Engelhard, 1967; Brückner, 1980; Hillebrecht, 1982; Jäger, 1984; Döppert, 1985, 1987; Kempf, 1985; Schenk, 1988, 1992a,b; Nitz, 1992) there are more or less detailed passages on the forest. Nitz (1983) deals, for example, with the role of the forest following Charlemagne's colonization, within a framework of technology history. Eberle (1990) describes the remains of log rafting (Keweloh, 1988) in the countryside of the Pfälzer Forest. Such studies continue the long tradition of historical–geographical preoccupation with the forest (Gradmann, 1901; Ziehen, 1970). The two decades following World War II were very productive in this respect: there was Mager's (1960) exemplary historical–geographical study on the forest in Old Prussia and Jäger's (1954, 1958, 1968) explanation of Central European forest development; other publications were, from a regional viewpoint, the collection about Schönbuch (Grees, 1969; Gamer-Wallert and Lorenz, 1999), and from a thematic viewpoint the studies by Hendinger (1960, 1966, 1967) which also made comparisons with the Scandinavian countries. On the other hand Tichy (1962) turned towards Italy.

Common to all these works is that the forest is never regarded as isolated from other spatial structures and elements. Geographical interest in the forest, in historically oriented geography, is thus as a rule also integrated into higher research questions (for example Fehn, 1982). This is significant because the main task of geography is the solution of problems with spatial relevance, not dealing with disciplinary theoretical questions in the sense of 'historical forest geography'. These considerations do not exclude calling such an approach to forests, for reasons of convention and linguistic simplification, 'forest and woods geography' or 'historical geography of forests and woods' or some such title. But it makes no sense to me to found a subdiscipline 'forest and woods geography' as long as there is no need for it. The relevance and possibilities of geography do not lie in a perfect division of the subject into highly

* Cf. yearly bibliographies in *Siedlungsforschung*, Bonn, with Central European production.

specialized subdisciplines but in the conversion – or at least the attempt at conversion – into synthetic scientific working.

2 Contributions of historical geography to forest history research

Historical geographers working in this field, therefore, should not focus on a 'theory of geography' but rather examine which specific contributions can solve spatial problems with its methods. Historical–geographical research, as I understand it, can develop its special strength best in a regional reference system[†] through the interpretation of archival and landscape sources by specification – i.e. verification, falsification or modification – of general knowledge on the origins of space. This means that the object of historical–geographical research related to the forest is not the forest itself but the processes and structures in the space determined by the extent of ecological and economic forest changes in the course of time. This includes, of course, detailed observation of the forest's condition and expanse as an element of the cultural landscape; under certain circumstances historical–geographical analysis includes spaces further away. The reconstruction of forest use and forest conditions in preindustrial times is thus primarily a historical–geographical regional structural analysis from a particular point of view, namely the use of forests.

Historical geography thus understood stands in context in generic regional research and is, furthermore, to be understood as part of general historical environment research (Pfister, 1985). Against the outlined general research background of genetic forest research, the importance of historical–geographical research of the 'forest' can be emphasized for the following fields:

1. It produces regionalized area balances for the distribution of woodland and open land (Darby, 1956; Bender, 1984; Busch *et al.*, 1985; Glaser *et al.*, 1991; Wulf and Schmidt, 1996). Here we are reminded of the German geographical classic by Schlüter and August (1959–1961), who first attempted an area balance for Central Europe. Their achievement cannot be rated high enough in spite of some criticism (Jäger, 1963). It does, however, require modification to include recent results. Modern tools like geographic information systems (GIS) are very helpful in this context (Schenk, 1993; Bund, 1998). Only such calculations allow well-founded statements on the relationship of forest–man–environment and the influence of the forest on historical events (Liebscher *et al.*, 1988; Gerlach, 1990), regional water balance, development of soils under forest (Ziehen, 1991), ground erosion (Müller, 1977; Bork, 1988;

[†] Hütteroth (1989) complains about the too regional approach of German historical geography and demands a less European-centred view.

Hahn, 1992) and accumulation (Nitz, 1989) as well as regional climate history (Pfister *et al.*, 1988; Glaser and Hagedorn, 1991). Research of this kind delivers basic information on regionalized energy (Baumgartner and Kirchner, 1980; Mizgajski, 1990; Kandler, 1992) and substance balances (Ulrich, 1989).

2. Such approaches should be closely connected with research on the changes of the inner structure of the forests (Schenk, 1992b) with the aim of obtaining a regional balance of forest conditions in space and time, as the composition and condition of the vegetation are parameters for the above-mentioned processes (Zeidler, 1970; Wilmanns and Bogenrieder, 1987; Wulf, 1997) as well as their result. It is thus necessary to include historical forest use methods, as Rubner (1960) and Bürgi (1998) did for the coppice-with-standards, in order to develop a visual representation of the condition of the forest at any particular time. Just how necessary this is, is clearly shown in the difficulties of a museum–didactic presentation of old landscape conditions in the form of exhibitions (Schenk, 1990; Schenk *et al.*, 1996).

3. Quantitative and qualitative changes in the structure of the forest should not be treated isolated from the social, economic and political conditions of their users. The transition to other forest uses influenced permanently the economic development potential and thus the social and demographic development of a region or vice versa (Eichhorn, 1952; Timm, 1960; Kampfmann, 1987, 1988; Mantel, 1987; Schenk, 1992a,b,c). The propagation of forest law (Radkau, 1983, 1986) and its realization in silvicultural practice depends to a great extent on the effectiveness of silvicultural administration, the ownership conditions and the socioeconomic conditions in the area of demand around a forest. Historical forest research must therefore take into consideration the regional geography of a certain area under the influence of economic and legal prerequisites typical of that time (Glaser *et al.*, 1988; Schenk, 1996b).

4. Without any doubt there is a lack of more extensive research (Brandl, 1987; Schenk 1996a) using quantifiable numbers (e.g. acorn feeding, financial proceeds, felling registers) as indicators for historic forest use and thus for forest conditions, as Mantel has shown with emphasis on the 'statistical period' from the early 19th century. Data processing is the key. Firstly, by using this method it is possible to speed up the usual cross-section method of observation and representation. Secondly, such investigation affords insight into the possible economic basis for the choice of certain area uses as well as use methods (Bürgi, 1999). Thirdly, the quantifying approach gives data for objectified comparisons of the economic, natural (Hausrath, 1907) and energy yields of the forests (including all these aspects see Schenk, 1999).

5. A multitude of literature on silviculture deals only with an area of forest which is limited according to present-day administrative requirements. The areas surveyed are often very small. This gives the works a regional depth but loses the wider context. Comparative sectoral works are rare (for example, Hessisches Ministerium fur Landwirtschaft, Forsten und Naturschutz, 1988), but it is these works that produce particularly important results which can not be obtained in any other way. Only thus can the causes of certain conditions or processes be determined in an exclusion or analogue method. Consequently we will see whether regularities and development tendencies of Central European forest history (cf. development phases by Mantel, 1971, 1973, 1975) outlined by some authors (Pacher, 1975) are also valid under spatial aspects (Schröder-Lembke, 1987) and can be fitted into a model of distribution and reconstruction of forests in 'non-sustainable times and regions' (Ritter, 1982). Principally it must be remembered that the division into epochs of forest history (IUFRO, 1973) is only transferable in a limited way to a geographical approach. If geography were to lean towards all these guidelines, then the spatial relationship of geographical questions and the specific possibilities of historical–geographical work would be neglected at the cost of historical–administrative, biographical, historical possession and silvitechnical aspects.[‡] Helpful as they are for the first approach by geographers to the theme 'development of forests', they refer primarily to the advance of silvicultural economy and its teaching and explain from that angle the condition of the forests – if they go into the spatial relationship at all. Thus, periodization and epoch formations come about which can be separated from general cultural landscape development, or even from the actual changes in the forest. The persistence of such area phenomena, which a forest represents, leads to rougher periodization of cultural landscape development (Born, 1974; Wiese and Zils, 1986; Jäger, 1987) as it results from the analysis of political, social and economic systems or from the development of sciences.

6. Moreover, a not insignificant number of silvihistorical works – after a cursory survey of forest history in early modern times based on a mostly superficial study of sources – start off with a detailed description of the first reliable silvicultural institutions at the turn of the 19th century or later. This temporal limitation can be justified partly by the age structure of many of today's forests, but eliminates traditional lines in many other questions and does not allow examination of important consistencies and breaks in its causes and stimuli for the present day. It would be better to trace back using 'source' silvigeographical works with a generic approach in the sense of a 'past-in-the-present' concept to early modern times and to pursue single indicators over as long a period of time as possible.

If historical geography deals in the above manner with forests – and most German historical geographers do so – then it achieves, by regional specification of general knowledge and the availability of regional findings for deducing general statements, indispensable contributions to general geography, namely to generically inclined economic geography, to regional

[‡] A typical example for this biographic–administrative forest history is Kremser (1990) and a survey of development of the German forest organization is to be found in Zundel (1979).

geography and to various disciplines of environmental history (Ebel and Ebel, 1988). It may also be that spatial planning (Hildebrandt, 1994; Schaal, 1994) and forest policy, by taking heritage into consideration, would find valuable information in the findings of historical geography.

References

Baumgartner, A. and Kirchner, M. (1980) Impacts due to deforestation. In: Bach, W. *et al.* (eds) *Interactions of Energy and Climate.* Dordrecht, pp. 305–316.

Bender, R.J. (1984) Die Entwicklung von Wald und Offenland im südlichen Pfälzer Wald. *Südwestdeutsche Schriften* 1, 93–103.

Bork, H.-R. (1988) Bodenerosion und Umwelt. Verlauf, Ursachen und Folgen der mittelalterlichen und neuzeitlichen Bodenerosion. Bodenerosionsprozesse, Modelle und Simulationen. *Landschaftsgenese und Landschaftsökologie* 13.

Born, M. (1974) *Die Entwicklung der deutschen Agrarland.* Darmstadt.

Brandl, H. (1987) Zur Geschichte der Wirtschaftlichkeit in der Forstwirtschaft. *Allgemeine Forstzeitung* 42, 1019–1023.

Brückner, H. (1980) Die Entwicklung der Wälder des Schwarzwaldes durch die Nutzung vergangener Jahrhunderte und ihre heutige Bedeutung. In: Liehl, E. and Sick, W.D. (eds) Der Schwarzwald. *Veröffentlichungen des Alemannischen Instituts* 47, 155–180.

Bund, B. (1998) Der Wandel der Kulturlandschaft Nordschwarzwald seit der 2. Hälfte des 19. Jahrhunderts. Eine historische Raum-Zeit-Analyse mit Hilfe eines Geographischen Informations Systems (GIS). Freiburg i. Br.

Bürgi, M. (1998) Waldentwicklung im 19. und 20. Jahrhundert. Veränderung in der Nutzung und Bewirtschaftung des Waldes und seiner Eigenschaften als Habitat am Beispiel der öffentlichen Waldungen im Zürcher Unter- und Weinland. *Beihefte Schweizerische Zeitschrift für Forstwesen* 84.

Bürgi, M. (1999) A case study of forest change in the Swiss lowlands. *Landscape Ecology* 14, 567–575.

Busch, U., Löffler, G. *et al.* (1985) Bodennutzung um 1820 und 1980. In: *Geschichtlicher Atlas der Rheinlande Beihefte* Vol. IV, 4.1–4.3. Köln.

Darby, H.C. (1956) The clearing of the woodland in Europe. In: Thomas, W.C. (ed.) *Man's Role in Changing the Face of the Earth.* Chicago, pp. 183–216.

Döppert, M. (1985) Die Waldentwicklung im Spiegel der frühneuzeitlichen Gesetzgebung, dargestellt am Beispiel des Schlitzer Landes. In: Nitz, H.-J. (ed.) *The Medieval and Early-Modern Rural Landscape*, pp. 231–241.

Döppert, M. (1987) Die Entwicklung der ländlichen Kulturlandschaft in der ehemaligen Grafschaft Schlitz unter besonderer Berücksichtigung der Landnutzungsformen. *Mainzer Geographische Studien* 29.

Ebel, F. and Ebel, I. (1988) Bibliographie zur Umweltgeschichte. *Umweltbundesamt Texte* 4.

Eberle, I. (1990) Holztrift im Pfälzer Wald. Der Finster-Breitenbach bei Elmstein. In: Custodis, P.-G. (ed.) *Technische Denkmäler in Rheinland-Pfalz. Spuren der Industrie- und Technikgeschichte.* Koblenz, pp. 141–143.

Eichhorn, O. (1952) *Der waldbauliche Gehalt der südwestdeutschen Forstordnungen.* Freiburg i. Br.

Engelhard, K. (1967) Die Entwicklung der Kulturlandschaft des nördlichen Waldeck seit dem Mittelalter. *Giessener Geographische Schriften* 10.

Fehn, K. (1982) Wirtschaftsentwicklung und Umweltbeeinflussung in Mitteleuropa aus historisch-geographischer Sicht (14. Jh.–20. Jh). In: Kellenbenz, H. (ed.) *Wirtschaftsentwicklung und Umweltbeeinflussung (14. Jh.–20. Jh.)*. Wiesbaden, pp. 277–292.

Gamer-Wallert, I. and Lorenz, S. (eds) (1999) *Der Schönbuch. Mensch und Wald in Geschichte und Gegenwart.* Tubingen.

Gerlach, R. (1990) Flußdynamik des Mains unter dem Einfluß des Menschen seit dem Spätmittelalter. *Forschungen zur Deutschen Landeskunde* 234.

Glaser, R. and Hagedorn, H. (1991) The climate of Lower Franconia since 1500. *Theoretical and Applied Climatology* 43, 101–104.

Glaser, R., Schenk, W. and Hahn, H.-U. (1988) Einflußgrößen auf die Anbau- und Ertragsverhältnisse des Ackerlandes im frühneuzeitlichen Mainfranken – Forschungsstand, Ergebnisse und offene Fragen. *Mainfränkisches Jahrbuch* 40, 43–69.

Glaser, R., Saurer, H. and Schenk, W. (1991) Abschätzung von Albedoänderungen in historischer Zeit und ihre Auswirkungen auf den Strahlungshaushalt im mainfränkischen Raum. *Meteorologische Rundschau* 44, 152–156.

Gradmann, R. (1901) Das mitteleuropäische Landschaftsbild nach seiner geschichtlichen Entwicklung. *Geographische Zeitschrift* 7, 361–435.

Grees, H. (ed.) (1969) *Der Schönbuch. Beiträge zu seiner landeskundlichen Erforschung.* Bühl and Baden.

Hahn, H.-U. (1992) Die morphogenetische Wirksamkeit historischer Niederschläge. Die Besselbergäcker und die Grünbachau – ein Beispiel aus dem Taubereinzugsgebiet. *Würzburger Geographische Arbeiten* 2.

Hausrath, H. (1907) Holzpreise, Holzhandelspolitik und Walderträge früher Zeiten. *Allgemeine Forst- und Jagdzeitung* 83, 333–339 and 369–375.

Hendinger, H. (1960) Der Wandel der mittel- und nordeuropäischen Waldlandschaft durch die Entwicklung der Forstwirtschaft im industriellen Zeitalter. *Geografiska Annaler* 42, 294–305.

Hendinger, H. (1966) Wandlungen der Waldbesitzstruktur und der Forstwirtschaft Frankens durch machtpolitische und wirtschaftsliberale Einflüsse in napoleonischer Zeit. *Forstwissenschaftliches Centralblatt* 85, 98–107.

Hendinger, H. (1967) Niederwaldwirtschaft und Lohrindengewinnung in Franken im 18. und 19. Jh. In: *Veröffertlichung XIV. IUFRO-Kongreß, Section 02.* Munich, pp. 168–179.

Hessisches Ministerium für Landwirtschaft, Forsten und Naturschutz (ed.) (1988) *Wald in Hessen, gestern – heute – morgen.* Mitteilungen der Hessischen Landesforstverwaltung No. 22.

Hildebrandt, H. (1994) Hachenburger Beiträge zur Angewandten Historischen Geographie. *Mainzer Geographische Studien* 39.

Hillebrecht, M.-L. (1982) Die Relikte der Holzkohlewirtschaft als Indikatoren für Waldnutzung und Waldentwicklung. *Göttinger Geographische Abhandlungen* 79.

Hütteroth, W. (1989) Quellen und Methoden historisch-siedlungsgeographischer Forschung außerhalb Europas. In: Denecke, D. and Fehn, K. (eds) *Geographie in der Geschichte.* Stuttgart, pp. 23–37.

IUFRO (International Union of Forestry Research Organizations) Subject Group 6.07 (ed.) (1973) *Leitfaden für die Bewertung von Regionalwaldgeschichten, Reviergeschichten und Bestandsgeschichten.* Zürich.

Jäger, H. (1954) Die Entstehung der heutigen großen Forsten in Deutschland. *Berichte zur Deutschen Landeskunde* 13, 156–171.
Jäger, H. (1958) Entwicklungsperioden agrarer Siedlungsgebiete im mittleren Westdeutschland seit dem frühen 13 Jh. *Würzburger Geographische Arbeiten* 6.
Jäger, H. (1963) Zur Geschichte der deutschen Kulturlandschaften. *Geographische Zeitschift.* 51, 90–143.
Jäger, H. (1968) Der Wald im nördlichen Süddeutschland in seiner historisch-geographischen Bedingtheit. *Mélanges de Géographie* 1, 597–613.
Jäger, H. (1984) Wüstungen, Forstbildung und Forstnutzung auf der Mainfränkischen Platte westlich von Würzburg. In: Jäger and Scherzer (eds) *Territorienbildung, Forsthoheit und Wustungsbewegung im Waldgebiet westlich von Würzburg (= Mainfränkische Studien 29).* Würzburg, pp. 11–78.
Jäger, H. (1987) *Entwicklungsprobleme europäischer Kulturlandschaften.* Darmstadt.
Kampfmann, G. (1987) Die Bevölkerungsentwicklung des Spessarts im Spiegel Kurmainzischer Forstordnungen. *Forstwissenschaftliches Centralblatt* 106, 294–299.
Kampfmann, G. (1988) Verkehrsgeographische und landeskundliche Aspekte der vorindustriellen Glasproduktion im Spessart. In: Kampfmann, G. and Krimm, S. (eds) *Verkehrsgeographie und Standorttypologie der Glashütten im Spessart.* Studien zur Geschichte des Spessartglases No. 2, pp. 9–109.
Kandler, O. (1992) Historical declines and diebacks of central European and present conditions. *Environmental Toxicology and Chemistry* 11, 1077–1093.
Kempf, A. (1985) Waldveränderungen als Kulturlandschaftswandel (Walliser Rhonetal). *Basler Beiträge zur Geographie* 31.
Keweloh, H.W. (1988) *Auf den Spuren der Flößer. Wirtschaftsund Sozialgeschichte eines Gewerbes.* Stuttgart.
Klohn, W. (1986) Verbreitung, Struktur und Funktion der Kleinstwälder und Gehölze in der Dümmerniederung und ihren Randbereichen – eine forstgeographische Untersuchung. *Vechtaer Arbeiten zur Geographie und Regionalwissenschaft* 3.
Klohn, W. and Windhorst, H.-W. (1984) Die Entwicklung der Forstwirtschaft in der Bundesrepublik Deutschland. *Zeitschrift für Agrargeographie* 2, 71–93.
Kremser, W. (1990) *Niedersächsische Forstgeschichte.* Rotenburg a.d. Wümme.
Liebscher, H.-J., Krahé, P. and Witte, W. (1988) Rekonstruktion der Witterungsverhältnisse im Mittelrheingebiet von 1000 n. Chr. bis heute anhand historisch hydrologischer Ergebnisse. *Abschlußbericht zum BMFT–Forschungsvorhaben LOF* 10/85.
Mager, F. (1960) *Der Wald in Altpreußen.* Köln and Graz.
Mantel, K. (1971) Entwicklungslinien der Forstwirtschaftslehre vom Mittelalter bis zum 19. Jh. In: Schwenk, S. (ed.) *Et Multum et Multa.* Festschrift für K. Lindner, Berlin and New York, pp. 201–240.
Mantel, K. (1973) IUFRO Subject Group 6.07 (ed.) Leitfaden für die Bewertung von Regionalwaldgeschichten, Reviergeschichten und Bestandsgeschichten, Zürich.
Mantel, K. (1975) Perioden der deutschen Forstgeschichte. *Allgemeine Forst- und Jagdzeitung* 146, 229.
Mantel, K. (1987) Die Anfänge der Waldpflege und Forstkultur im Mittelalter unter der Einwirkung der lokalen Waldungen in Deutschland. *Forstwissenschaftliches Centralblatt* 87, 75–100.
Mizgajski, A. (1990) Entwicklung von Agrarlandschaften im Mitteleuropäischen Tiefland seit dem 19. Jh. in energetischer Sicht. *Münstersche Geographische Arbeiten* 33.

Müller, S. (1977) Waldböden als Maßstab der Bodenerosion in Baden-Württemberg. *Jahrbuch Geologisches Landesamt Baden-Württemberg* 19, 129–141.
Nitz, H.-J. (1983) Feudal woodland colonization as a strategy of the Carolingian Empire in the conquest of Saxony. In: Roberts, B.K. and Glasscock, R.E. (eds) *Villages, Fields and Frontiers. BAR International Series* 185, pp. 171–184.
Nitz, H.-J. (1989) Strukturen der historischen Kulturlandschaft in ökologischer Perspektive. In: Herrmann, B. and Budde, A. (eds) *Naturwissenschaftliche und historische Beiträge zu einer ökologischen Grundbildung.* Hannover, pp. 189–198.
Nitz, H.-J. (1992) Historical geography. In: Ehlers, E. (ed.) *40 Years after German Geography.* Bonn and Tübingen, pp. 145–172.
Pacher, J. (1975) Entwicklungstendenzen der Forstwirtschaft in Deutschland in der ersten Hälfte des 19. Jahrhunderts unter dem Einfluß allgemeiner Zeitströmungen. *Allgemeine Forst- und Jagdzeitung* 146, 111–120.
Pfister, C. (1985) *Klimageschichte der Schweiz 1525–1860*, 2nd edn. Bern and Stuttgart.
Pfister, C. *et al.* (1988) *Witterungsextreme und Waldschäden in der Schweiz.* Bern.
Ritter, W. (1982) Wirtschaftsentwicklung und Umweltbeeinflussung (14 Jh.–20. Jh.). In: Kellenbenz, H. (ed.). Wiesbaden, pp. 89–103.
Röll, W. (1966) Die kulturlandschaftliche Entwicklung des Fuldaer Landes seit der Frühneuzeit. *Giessener Geographische Schriften* 9.
Rubner, H. (1960) Die Hainbuche in Mitteleuropa. Untersuchungen über ihre ursprünglichen Standorte und ihre Förderung durch die Mittelwaldwirtschaft. *Forschungen zur Deutschen Landeskunde* 121.
Rubner, K. (1928) Der Wald als geographische Erscheinung. *Tharandter Forstliches Jahrbuch*, 399–411.
Schaal, R. (1994) Waldgeschichtliche Erhebungen im Forstbezirk Münsingen als Beitrag zur Waldbauplanung. *Mitteilungen des Vereins für Forstliche Standortkunde und Forstplanzenzüchtung* 37, 61–65.
Schenk, W. (1988) Mainfränkische Kulturlandschaft unter klösterlicher Herrschaft. Die Zisterzienserabtei Ebrach als raumwirksame Institution vom 16. Jh. bis 1803. *Würzburger Geographische Arbeiten* 71.
Schenk, W. (1990) Vom Kloster zum Dorf: Kulturlandschaftswandel auf der Gemarkung des ehemaligen Zisterzienserinnenklosters Frauental. Exemplarisches und Besonderheiten zur Landschaftsgeschichte Frankens seit 1500. *Württembergisch Franken* 74, 121–161.
Schenk, W. (1992a) Waldnutzung, Waldzustand und regionale Entwicklung in vorindustrieller Zeit im mittleren Deutschland: historisch-geographische Beiträge zur Erforschung von Kulturlandschaften in Mainfranken und Nordhessen. *Erdkundliches Wissen* 117.
Schenk, W. (1992b) Nutzung und Zustand der Gemeinschaftlichen Waldungen in der Grafschaft Wertheim am Ende des 18. Jahrhunderts – ein Beitrag zur Kulturlandschaftsgeschichte Frankens. *Wertheimer Jahrbuch* 1991/1992, 97–119.
Schenk, W. (1992c) Fundamental changes in the forest landscapes of Lower Franconia (Germany) in the nineteenth century. In: Verhoeve, A. and Vervloet, J.A.J. (eds) *The Transformation of the European Rural Landscape. Methodological Issues and Agrarian Change 1770–1914.* Brussels, pp. 249–258.
Schenk, W. (1993) The use of CAD and GIS systems in the reconstruction of large-scale historic field systems and land utilisation – an example out of the sixteenth century in southern Germany. *History and Computing* 5(1), 25–34.

Schenk, W. (1996a) Forest development types in Central Germany in pre-industrial times. A contribution by historical geography to the solution of a forest history research argument about the 'wood scarcity' in the eighteenth century. In: Cavaciocchi, S. (ed.) *L'uomo e la foresta secc. XIII–XVIII.* (Series II: Atti delle 'Settimane di Studi' e altri Convegni 27.) Prato, 201–223.

Schenk, W. (1996b) Nebennutzungen im Spiegel von herrschaftlichen Forstrechnungen des 17.–19. Jahrhunderts aus Franken – zugleich ein Diskussionsbeitrag zum 'Aufbau Langer Reihen' für die Umweltgeschichte. *News of Forest History* 25/26, 98–110.

Schenk, W. (ed.) (1999) *Aufbau und Auswertung 'Langer Reihen' zur Eforschung von historischen Waldzüstanden und Waldentwicklungen (Tübinger Geographische Studien 125).* Tübingen.

Schenk, W., Stöhr, L. and Layer, G. (1996) Volkskunde und Geographie. Zur Zusammenarbeit von Volkskundlern und Geographen in Museums- und Ausstellungsprojekten in Mainfranken. *Bayerische Blätter für Volkskunde* 23(1), 21–28.

Schlüter, O. and August, O. (1959–1961) *Atlas des Saale- und mittleren Elbegebietes.* Parts 1–3. Leipzig.

Schröder-Lembke, G. (1987) Waldzerstörung und Walderneuerung in der vorindustriellen Zeit. *Zeitschrift für Agrargeschichte und Agrarsoziologie* 35, 120–137.

Tichy, F. (1962) Die Wälder der Basilicata und die Entwaldung im 19. Jh. Vorgänge, Ursachen und Folgen. *Heidelberger Geographische Arbeiten* 8.

Timm, A. (1960) *Die Waldnutzung in Nordwestdeutschland im Spiegel der Weistümer.* Köln.

Troll, C. (1962) Die Stellung des Waldes in den deutschen Kultur- und Wirtschaftslandschaften. *Schriften des Deutschen Forstwirtschaftsrates,* 3–30.

Ulrich, B. (1989) Die historische Entwicklung des Beziehungsgefüges Wald–Mensch–Umwelt. In: Herrmann, B. and Budde, A. (eds) *Naturwissenschaftliche und historische Beiträge zu einer ökologischen Grundbildung.* Hannover, pp. 95–106.

Wiese, B. and Zils, N. (1986) *Deutsche Kulturlandschaften. Wandel und Bewahrung deutscher Kulturlandschaften.* Herford.

Wilmanns, O. and Bogenrieder, A. (1987) Zur Nachweisbarkeit und Interpretation von Vegetationsveränderungen. *Verhandlungen der Gessellschaft für Ökologie* 16, 35–44.

Windhorst, H.-W. (1978a) *Geographie der Wald- und Forstwirtschaft.* Stuttgart.

Windhorst, H.-W. (ed.) (1978b) *Beiträge zur Geographie der Forst- und Waldwirtschaft (Wege der Forschung 517).* Darmstadt, pp. 517.

Wulf, M. (1997) Plant species as indicators of ancient woodland in northwestern Germany. *Journal of Vegetation Science* 8, 635–642.

Wulf, M. and Schmidt, R. (1996) Die Entwicklung der Waldverteilung in Brandenburg in Beziehung zu den naturräumlichen Bedingungen. *Beiträge für Forstwirtschaft und Landschaftsökologie* 30(3), 125–131.

Zeidler, H. (1970) Edaphisch und anthropogen bedingtes Vegetationsmosaik in Wäldern. *Gesellschaftsmorphologie* 322–333.

Ziehen, W. (1970) Wald und Steppe in Rheinhessen. Ein Beitrag zur Geschichte der Naturlandschaft. *Forschungen zur Deutschen Landeskunde* 196.

Ziehen, W. (1991) 350 Millionen Jahre Waldböden. *Allgemeine Forst- und Jagdzeitung* 162, 154–156.

Zundel, R. (1979) Die Entwicklung der Forstorganisation der deuschen Länder in den letzten 150 Jahren. *Allgemeine Forst- und Jagdzeitung* 150, 9–22.

Methods Towards Studying Historical Changes in Forest and Landscape Patterns: a Comparison between Two Nearby Socioeconomic Contexts

12

V. Robiglio

Department of Forestry and Environmental Sciences, Università di Firenze, Via San Bonaventura 13, 50145 Florence, Italy

This chapter compares the landscape pattern dynamic of two alpine valleys. We used a hierarchical framework to analyse land use and farming systems and to fit these into the landscape system hierarchy. We considered different levels of the spatial and temporal scale in order to assess the influence of land use as a landscape-structuring factor. Historical cadastres, maps, written documents, statistics and aerial photographs were used. The system of production and the farming structure were shown to be significant driving forces in landscape organization. To understand current ecological dynamics, landscape systems should be viewed in a historical perspective.

1 Introduction

Landscapes in alpine areas are currently altered by long-term global climate changes and short-term land use changes (extensification–abandonment versus intensification and development of tourist infrastructure). A recent workshop (Intecol, Florence in July 1998) discussed the effects of land use changes on landscape patterns and dynamics in mountainous areas; the need to consider spatial and temporal scales in ecological analyses of landscapes and the connection to the socioeconomic and institutional circumstances were put forward.

The development of the spatial configuration of an anthropic landscape is produced by the complex interactions between physical, biological and socio-economic forces in time, which are specific for individual regions (Forman and Godron, 1986). Changes in the landscape are due to changes in management practices and in the sociopolitical and economic forces controlling land use. In the present chapter, the way landscape dynamics are linked to the changes of land use is analysed by comparing the development of the landscape systems of two contiguous valleys, located in similar environments, but inhabited by two different populations.

Two different societies, even if surrounded by the same environment, may produce agricultural systems with different features and infrastructures, such as the structure of land property, the type and size of the productive plots (edges, fences, limits), and the system of management and of communication lines (roads, paths, edges, etc.).

The analysis of the differences in space and time between the landscape systems aims at: (i) isolating the physiographical and human impact on the alpine cultural landscapes, in order to assess the role of land use as a pattern-structuring force; (ii) identifying the mechanisms responsible for landscape structure dynamics at the different levels of the landscape system; and (iii) identifying the spatial temporal scale at which each mechanism is significant.

Landscape patterns were analysed according to the time ranges, defined by frequency, intensity and qualification of available historical resources. Starting from 1759, historical cadastres and maps were used to trace back the landscape patterns at the field level. Concepts derived from landscape ecology and land evaluation were applied to the historical reconstruction and to the results of landscape system analysis.

2 Study site

The Mastallone valley is located on the Italian side of the western Alps, on the border between the wide hydrological basins of the Sesia river and the Verbano (Fig. 12.1). The upper part of the Mastallone Basin (6000 ha) consists of two main deep V-shaped valleys of glacial origins, successively hollowed by the action of the Mastallone stream and its main tributary, the Landwasser stream (confluence 774 m a.s.l.). Striking differences occur in the morphology of the west- and east-exposed slopes, the latter being steeper. Altitude ranges from 770 to 2450 m a.s.l. The morphological differences between the west- and the east-exposed slopes are reflected by land use and vegetation cover: more-exposed sites are still cultivated, whereas shrubs or forests cover the enclosed and shady zones.

According to the physiognomic division of vegetation into altitudinal belts, the Mastallone Basin is included in the 'piano montano' and in the 'piano

Fig. 12.1. Location of the study area.

culminale', distinguished in the 'orizzonte subalpino' (crabbed shrubs) and in the 'orizzonte alpino' (pastures). Three main groups of vegetation cover may be distinguished: grasslands, shrublands and forests. The vegetation of the grassland and pastures is pioneer-like and floristically poor on screes and rocks. It becomes more rich and complex at lower altitude and on less disturbed surfaces.

The meadows are subdivided into still manured meadows (near the villages), mowed meadows and abandoned meadows invaded by hazels, ash trees, maples and raspberry bushes. In the shrubberies (1200–2000 m a.s.l.), *Alnus viridis* prevails with some scattered *Betula pendula* trees.

The main occurring forest types are fir woods and beech woods, which are frequently contiguous or intermixed, located at an altitude between 1000 and 1500 m a.s.l. Occurring species are *Sorbus* sp., *Laburnum* sp., *Betula pendula, Acer pseudoplatanus* and, on the upper border, *Larix decidua. Picea abies* is rare. Rare oak forests (*Quercus robur*) occur at a lower altitude (up to 1100 m), on south-exposed slopes and in relatively open stands. Some scattered larch (*Larix decidua*) forests occur on the upper part of the slopes. Alder and *Rhododendron* prevail in the undergrowth. Pioneer broad-leaf forests with *Betula pendula, Populus tremula, Laburnum* sp., etc. are located in rocky areas or on screes. In short and narrow valleys and on the small *impluvia*, a pioneer vegetation of *Alnus* sp. occurs, which evolves in a complex with *Fraxinus excelsior, Acer pseudoplatanus, Ulmus glabra* and *Tilia* sp.

2.1 Land use development of the valleys of Rimella and Fobello

Since the 13th century, the upper part of the Landwasser valley has been colonized by groups of the Walser population. Originally from the Vallese, the Walser were highly specialized shepherds, looking for new pastures and land; starting from the head of the valley, they colonized the area rapidly. The settlement in the contiguous valley of Fobello occurred later and gradually, shepherds came from the ValSesia and Po plains for the seasonal mountain pastures.

In both territories, the *Alpwirtschaft* model (alpine mixed agricultural model) developed – a model characterized by the fragile balance between the main productive functions of agriculture (self-consuming) and cattle breeding.

3 Methods

3.1 Land use

A 'systematically connected' way to evaluate land use was found in the land evaluation methodology, providing means for combining the ecological and technical aspects of land use with economic and social circumstances. For an ecological interpretation, land use needs to be classified in categories of ecologically significant factors bringing about changes in the landscape of concern. For the area of interest an ecologically relevant land use classification scheme has been developed on the basis of the *Field Guide for the Description of Terrestrial Landscape Ecological Sample Unit* (Stortelder and Vos, 1983).

In land evaluation the concept of land utilization type (LUT) (Beek, 1978) is used to classify land use by combining its ecological and technical aspects with economic and social conditions. A LUT is defined as a coherent complex of human activities which are applied over a long period by man organized in coherent systems on any given tract of land. The decisional unit (coherent system) is a natural unit, like a family farm. The data required to describe a LUT are various and depend on the purpose (e.g. agriculture, forestry, conservation, etc.). Attributes relevant for an ecological typification of a LUT are produce, labour, technology and management.

The combination of a given LUT with a specific land unit is called a land utilization system (LUS) which is a technological and ecological entity (Vink, 1983). The LUS is a subsystem of a farm system, defined as a complex of relationships between land, labour, capital and socioeconomic factors directed by an entrepreneur on the basis of his knowledge, goals and methods. Furthermore, these aspects are investigated by the farming system analysis (FSA) which, paying attention to the multiple factors that govern the functioning of a farm system and according to agroecological and socioeconomic criteria, consider the production patterns spatially distributed within a given area. The theoretical base of the FSA derives from the analogy between agriculture and ecology, which introduces the concept of agriculture as a hierarchy of systems. Within the agricultural hierarchy of systems, the farm system is part of a higher system at the subregional level. This may be a socioeconomic unit, (e.g. a village) or a physical unit (e.g. a valley or a watershed), depending on the perspective of analysis. The farm system, in turn, is a decisionmaking and land use unit, which comprises a complex of subsystems: the farm household and cropping and livestock systems. Each of these is a complex of subsystems by itself.

The selected land utilization characteristics with ecological side effects, produce, technology and management, were described by recording field indicators, integrated by information obtained through the agricultural inquiries and statistics delivered by the municipalities and the *comunità montana*. A classification system of the local main LUT was developed to facilitate the integration with other parameters and the comparison of different sytems.

3.2 Vegetation as a diagnostic criterion in the landscape

Landscape changes are strongly expressed by the changing of the vegetation cover, either due to a natural succession or by human intervention. In this research vegetation is interpreted as a *diagnostic criterion* in the landscape. It has a high descriptive capacity and an indicative value due to the large number of associated attributes of the landscape with relevance for plant communities; furthermore, it provides important clues to past land use processes and their consequences. On the western side of the Alps the natural regeneration of woodland in abandoned fields is a massive phenomenon. The way secondary succession develops is partially dependent on the previous land use as was analysed in several studies (Salbitano, 1993).

3.3 Methods

Landscape dynamics are analysed at different spatial scales; the concept derived from landscape ecology was combined with detailed analysis on vegetation community and renovation of forest trees in pilot areas. Two main scale levels have been considered: the scale of landscape patterns and the scale of land units.

Aerial photo interpretation

Landscape patterns were first mapped by a preliminary interpretation of true colour aerial photographs from 1976 (scale 1/13,000), with subsequent reconnaissance ground surveys. A land cover/land use legend was produced. The results, combined with the information from pre-existing thematic maps, were used to compile a land unit map at a scale of 1 : 25,000. Temporal changes in the landscape mosaics were assessed through a sequence of aerial photographs, using the developed legend to produce maps. The first aerial photographs of the area of the Alta Val Mastallone date from the 1950s;

successive photos come from the 1970s and the most recent ones are from the 1990s. Each photo was viewed with adjacent photography under a mirror stereoscope to produce stereoimages and magnification of the land cover. A transparent grid with cells representing 150 m^2 was then placed over the land use acetate sheet. Twenty land cover categories were used on the base of the developed legend (Table 12.1). Contiguous (horizontally and vertically) cells of the same land cover were considered a patch. Each patch was located in the landscape matrix and its size was calculated.

The output of this phase was a general qualitative analysis of the spatial patterns and the comparison of patterns through time.

Pattern analysis

Patterns were quantified by using landscape indices. For two areas of 185 ha, on the west-exposed slopes of the valleys, size and number of patches were counted for each temporal range. Indices based on information theory were used: (i) the H index, measuring diversity; and (ii) the D index, measuring dominance within a landscape pattern.

$$\mathrm{H} = -\sum_{i=1}^{m}(\mathrm{P}i)\ln(\mathrm{P}i) \quad \text{and} \quad \mathrm{D} = \ln(s) + \mathrm{H}$$

where Pi is the proportion of the landscape in cover type i; m is the number of cover types observed; and s is the maximum proportion of land cover when all land cover types are present in equal proportion. The larger the value of H the more diverse the landscape; the larger the value of D the more the landscape is dominated by one or few land cover types.

Field surveys

Based on the previous comparison, the west-exposed slopes were selected for a finer analysis at the slope level: stratified random sampling was done along transects following the slope sequence from the crest to the bottom. The size of the sample plots was connected to the concept of minimal area (Muller-Dombois and Ellenberg, 1974). Data on the abiotic and biotic factors and on land use were systematically collected using specific suitable forms. Structure and floristic composition were described according to the Zurich–Montpellier approach. The observation forms were organized in a database program. For the classification of the vegetation data, a divisive hierarchical clustering method was applied to establish community types.

The typification of vegetation types and LUT enabled us to focus on a more detailed analysis in 16 selected pilot areas (20 × 20 m). Plots were located in such a way that data on border and interior features of the

Table 12.1. Mastallone basin land cover classification.

	Category	Description
1.	A1	Grassland
2.	A2	Grassland: sparse trees; crown closure < 20% ground cover; isolated trees
3.	A3	Grassland: sparse trees; crown closure < 20% ground cover; trees in small groups
4.	A4	Grassland: sparse trees; crown closure < 50% ground cover
5.	A5	Grassland: shrubs; crown closure < 50% ground cover; irregular texture; height < 1.5 m
6.	A6	Grassland: woodland; crown closure < 50% ground cover; irregular texture
7.	B1	Low shrubland: height < 1.5 m; cover > 50%; trees small and isolated, if present < 20% ground cover
8.	B2	Low shrubland: height < 1.5 m; cover > 50%; trees small and isolated or in groups at between 20% and 50% ground cover
9.	C1	Woodland: > 50% ground cover; height < 4 m; grassland visible at < 20% ground cover
10.	C2	Woodland: > 50% ground cover; height < 4 m; taller trees at < 50% ground cover
11.	D1	Coniferous forest: canopy closure at > 80% ground cover
12.	E1	Mixed coniferous–broad-leaf forest: canopy closure at > 80% ground cover
13.	F-A	Mixed broad-leaf forest: high stand; canopy closure at < 70% ground cover; remaining area is covered by grassland
14.	F-B	Mixed broad-leaf forest: high stand; canopy closure at < 70% ground cover; remaining area is covered by scrubland
15.	F-ir	Mixed broad-leaf forest: canopy closure at > 70% ground cover; irregular structure (irregular texture)
16.	F-rg	Mixed broad-leaf forest: canopy closure at > 70% ground cover; regular structure (regular texture)
17.	G-A	Beech forest: high stand; canopy closure at < 70% ground cover; remaining area is covered by grassland
18.	G-B	Beech forest: high stand; canopy closure at < 70% ground cover; remaining area is covered by scrubland
19.	G-ir	Beech forest: high stand; canopy closure at > 70% ground cover; irregular structure (irregular texture)
20.	G-rg	Beech forest: high stand; canopy closure at > 70% ground cover; irregular structure (regular texture)

vegetation units were recorded in order to analyse the interaction between land cover types (e.g. successional dynamics). Floristic data were also collected according to the Zurich–Montpellier approach; for all the species an indicator value describing the ecological factors was reported

(Landolt, 1977). Considering the tree component, the stand structure was described – frequency, diameter, height and age were measured, and damage caused by grazing was noted. The relation between the parameters was plotted. Each site was mapped, and profiles and ground plans were drawn (tree position, crown projection).

A transect for each pilot area was traced and measurements of saplings and seedlings were done, aiming to identify a regeneration model and relating it to the developmental tendency of the succession process.

The taxonomic structure was examined by applying the dominance index of Simpson:

$$I = \sum_{i=1}^{S} (ni/N)^2$$

where ni is the value of one species for the considered parameter; N is the value of all species; and S is the species number.

The spatial distribution was examined by applying the dispersion index, which showed the model of spatial distribution: aggregate, uniform and random.

$$Id = \frac{\sum_{i=1}^{N} (x - x_m)^2 / N - 1}{\sum x/N}$$

where x is the number of individuals in an area; x_m is the mean number of individuals; and N is the sample plots number.

3.4 Historical sources and methods

Written historical sources were used to: (i) trace back the spatial and temporal dynamics of the settlements; and (ii) reconstruct previous forms of land use, land management practices and products including the dynamics of land ownership.

The 18th century valley landscape patterns were reconstructed by using a set of documents and technical land use map originally for mining purposes, the *Gran Carta della Valsesia* from 1759. The source classifies forest high-stands on the base of species and dimensional parameters. The 19th century cadastre – *Catasto Rabini* 1861 – and 20th century cadastre – *Nuovo Catasto Terreni* 1954 – permitted a trace to be made of the landscape systems at the field level and to follow them through the years. Census registers (starting from 1861) and agricultural inquiries and statistics (1931 and 1990) were used to study population dynamics and social changes during the late 19th and 20th centuries. Information from archives and written documents in the form of articles, books, etc. and personal communication from local people and historical photographs (1901) were gathered.

4 Results

According to the preliminary description in the cadastre of 1863, the agricultural systems developed in the Fobello and Rimella territories were based on one main LUT: cattle breeding. But the socioeconomic context of the systems differed significantly.

4.1 Rimella

The socioeconomic system in Rimella was closed, based only on cattle breeding and aiming at a total autharchy. All suitable areas around the villages were used for agriculture for food supply; the relevant crops were potatoes and hemp. The remaining parts were used for livestock breeding, and meadows for haymaking and pastures.

The Walser people of Rimella had very little contact with the neighbouring communities and the main centres of the valley bottom.

4.2 Fobello

The socioeconomic system of Fobello was based on livestock breeding and diversified productive activities: wood from beech and fir forests was systematically marketed and, since the beginning of the 20th century, tourist and commercial activities developed.

4.3 Emigration and socioeconomic changes during the 20th century

During the greater part of the last centuries migration was a 'safety valve' to maintain the balance between human load and land resources. In both territories surveyors recorded population movements. In Rimella migration was male and seasonal (summer), whereas migration in Fobello lasted years. Some migration of entire groups of families was permanent. Men of Fobello tended to migrate to the big centres of northern Italy and to the capitals of Europe; traditionally they were cooks, gastronomists and hotelkeepers.

Towards the turn of the 19th century a real exodus on a massive scale began (Fig. 12.2). In line with the striking and continuous decline of the number of residents, the farming system also changed. Three elements are striking:

1. the remarkable drop in cattle breeding;
2. the continuous increase of goat breeding;
3. a marked drop of the agricultural surface (*SAU*) in Rimella.

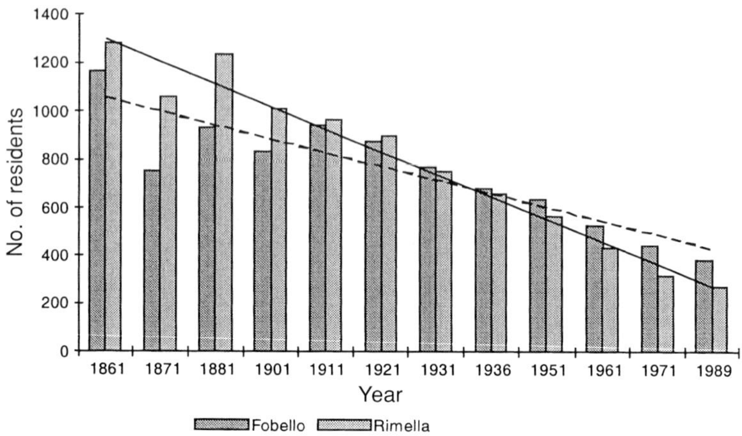

Fig. 12.2. Demographic dynamics in the municipalities of Fobello and Rimella.

Table 12.2. Municipalities land use distribution.

	Fobello	Rimella
Surface		
Farm's total surface (ha)	1383	1101.52
Utilized agricultural land (ha)	967.31	299.45
SAU% of total	69.80%	27.20%
Farm's surface productive organization		
Crops	0.07%	0.09%
Permanent meadows and pastures	69.00%	27.10%
Woodland	15.00%	10.20%
Other	15.30%	62.50%

The farm number decrease was also remarkable in Rimella. If livestock is considered in terms of density (number of head ha^{-1} of agricultural surface):

1. the herds of cattle decreased more drastically in Fobello (0.9 → 0.3);
2. the herds of goats increased remarkably in Rimella (0.6 → 1.25).

The main consequences are: (i) the abandonment of permanent and temporary settlements at higher altitudes; (ii) the contraction of the summer transhumance (seasonal grazing activity); (iii) the abandonment of remote pastures; (iv) the overexploitation of the closed fertile pastures; and (v) total abandonment of grazing on previous hayfields and crops (Table 12.2).

4.4 Changes in the landscape system (1950–1990)

The proportion of land in each cover type has varied since the 1960s. Agricultural land declined in both territories; forest cover increased in overall abundance; transitional land, initially increasing, decreased rapidly in Rimella, where broad-leaf mixed forest increased remarkably. Patch numbers decreased in both territories, significantly so in Rimella; and the richness index increased slightly. Indices of diversity exhibited the greatest net changes in Rimella, increasing between 50^S and 70^S and decreasing between 70^S and 90^S (2.1–2.4–2). Dominance indices did not differ significantly through time (0.7–0.5–0.6). Fobello showed stability.

The 'slope' level

In order to localize the dynamic phenomena, the slope level was considered. As a result of the cluster analysis of relèves, vegetation types were identified and located along the slopes (Fig. 12.3). The section of Fobello was characterized by the occurrence of *Abies alba* stands, *Fagus sylvatica* stands and mixed broad-leaf stands with a dominance of *Fagus sylvatica*. The section of Rimella was characterized by *Nardus stricta* and *Festuca ovina* grasslands, *Larix decidua* and *Picea abies* stands (plantations), and *Acer pseudoplatanus* and *Fraxinus excelsior* mixed broad-leaf stands. Further differences were: (i) the spatial distribution of the vegetation types along the slopes and their

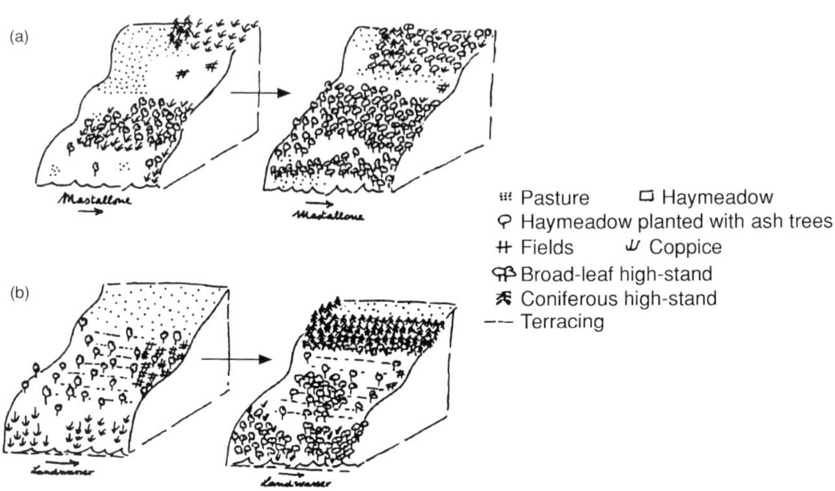

Fig. 12.3. Slope sketches 1863–1996: (a) Fobello–Mastallone; (b) Rimella–Landwasser.

interrelationships; and (ii) the structural differences occurring within the same vegetation type.

Land use management

In Rimella, land management was strictly organized. The reconstruction from historical sources shows a striking separation of settlements (concentrated along one side of the slope) and productive areas. Important land improvements were recorded in the cadastre of 1863: (i) the occurrence of stone walls and terraces (*campi a scaglioni, campi a muricci*); and (ii) the planting of ash trees in agricultural fields or meadows for fodder-leaf production (for hay integration). The slope was divided in wide productive belts: wide, open pastures and wide, planted meadows. The latter were complexes of small land lots, surrounded by edges of ash trees and maples and planted with isolated ash trees. Pastures were common land.

In Fobello the settlements were scattered; the organization of the productive areas was at a village level. There were farmer *consortia* managing high-forest (*Abies alba* and *Fagus sylvatica*) stands and pastures. Pastures were scattered in small lots along the slope and in many cases the cadastre of 1863 noted the occurrence of stones and bushes: *pascolo sassoso con cespugli* (stony and bushy pasture). There are no records of improvements.

Land unit level

At the land unit level the differences showed by the two systems were striking. Tables 12.3 and 12.4 summarize the result of the analysis in the 16 pilot areas. The sample plots were followed through the 19th century, according to the land use classifications of the 19th and 20th cadastre and the field surveys of 1996. In 1996 the plots differed in physionomy, structure and specific composition of the vegetation. *Fraxinus excelsior* and *Acer pseudoplatanus* proved to be absolutely dominant in the rapid successional processes occurring in the previous meadows and crop lands of Rimella. The dynamics observed in the sample plot 'meadows' of Fobello are slow, mainly changing to *Quercus petraea, Fraxinus excelsior, Fagus sylvatica, Corylus avellana.* In the previous pastures of Fobello a new woodland has developed.

Land use

The production of fodder leaves in the meadows planted with ashes in Rimella was important as the leaves were integrated into hay production. In the 1950s cadastre the classification of land uses is quite similar to the earlier one except for the disappearance of the tree plantations, which were probably not recorded because in 1956 they had already lost economic importance.

Table 12.3. Land use and cover in 16 pilot areas in the Fobello and Rimella Valleys.

1863	1956	1996	Recolonizing species
Crop	Pasture	Fobello: grassland recolonized by tree species	Fobello: *Fraxinus excelsior, Acer pseudoplatanus*
	Meadow	Rimella: new woodland, complex structure	Rimella: *Fraxinus excelsior, Acer pseudoplatanus*
	Crop		
Meadow	Meadow	Fobello: grassland recolonized by tree species	Fobello: *Quercus petraea, Fraxinus excelsior, Fagus sylvatica, Corylus avellana*
	Pasture	Rimella: new woodland, complex structure	Rimella: *Fraxinus excelsior, Acer pseudoplatanus*
Pasture	Pasture	Fobello: new woodland, complex structure	Fobello: *Betula pendula, Populus tremula, Fagus sylvatica, Laburnum* sp., *Sorbus aucuparia, Quercus petraea, Acer pseudoplatanus*
		Rimella: grassland recolonized by tree species	Rimella: *Betula pendula, Populus tremula, Fagus sylvatica, Laburnum* sp., *Sorbus aucuparia*

The intensity and cutting system for fodder-leaf gathering was reconstructed through interview with farmers. Whereas in Fobello ash trees were pruned or pollarded, in the Rimella area the crown was kept in shape to allow children and women to climb up and pick every single leaf. In order to keep the leaf production high only the unreachable branches were cut.

The present LUS was distinguished on the base of land use intensity, such as grazing species and grazing intensity. In Fobello, cattle and goat grazing is still widespread on pastures and meadows, and in bushy and wooded areas.

5 Discussion

Spatial patterns of landscape have changed since the 1950s. The landscape pattern differences decreased because of the increase in both territories of

Table 12.4. Quantitative parameters in the pilot areas.

Area	Nsp	dn m^{-2}	Gbcm^2 m^{-2}	hm (m)	I(n)	I(g)
F1	5	1.0	1.97	3.2	0.51	0.77
F2	6	5.2	2.65	1.5	0.36	0.05
F3	7	2.55	10.33	7.0	0.3	0.07
F4	11	3.73	7.4	3.4	0.14	0.19
F5	8	1.97	2.43	3.8	0.23	0.73
F6	8	1.98	5.69	1.6	0.35	0.37
F7	5	1.31	0.62	2.4	0.39	0.66
F8	4	3.2	6.96	3.6	0.14	0.75
R1	5	3.27	2.17	2.3	0.4	1.00
R2	4	0.82	2.83	5.7	0.7	0.99
R3	8	4.7	7.4	3.6	0.2	0.6
R4	2	4.6	1.95	1.4	0.5	0.5
R5	7	8.32	5.5	4.0	0.3	1.3
R6	2	6.5	4.1	3.4	0.3	0.3
R7	9	11.17	10.91	3.4	0.3	0.3
R8	2	12.78	42.2	5.7	0.5	0.8

Nsp = number of species;
Dn m^{-2} = numerical density (n m^{-2});
Gbcm^2 m^{-2} = basal area density (G m^{-2});
hm (m) = average height (m);
I(n) = Simpson's dominance index referred to the numerical density;
I(g) = Simpson's dominance index referred to the basal area density.

forest cover, leading to the simplification of the original patterns. The changes can be summarized as: (i) a decrease of grasslands; (ii) an enlargement of forested patch size; and (iii) a confluence of fine-scale scattered point and linear tree elements. The changes are related mainly to a clear decrease in land use in Rimella and to an extensification of land use in Fobello. The diminishing control of humans over agricultural ecosystems is linked to depopulation and social changes. The rapid changes in Rimella are linked to the collapse of the highly specialized Walser agricultural system, whereas in Fobello a slow extensification and marginalization of the traditional land use has occurred.

In both territories the dominant productive activity was the same: livestock breeding. Within a similar productive frame, differences arose in: (i) farming system – size, number and species of livestock; (ii) hay production; and (iii) land ownership and land management. The main aspects with direct consequences on the landscape pattern were: (i) crop field and haymeadow size; (ii) the type of combination of fields along the slope; and (iii) the intensity and method of fodder-leaf production from the 'cultivated' *Fraxinus excelsior* and *Acer pseudoplatanus* trees.

The dominance of *Fraxinus excelsior* and *Acer pseudoplatanus* in the recolonization dynamics in Rimella is strictly linked to the past fodder-leaf gathering activity determining a high density of mature trees 'ready to disseminate' in fallow meadows and abandoned crop fields. The small land lots (0.3 acres) were arranged in a fine-scale patchy pattern, with small open spaces (grasslands) and linear or point elements of isolated ash and maple trees or hedges. The massive and rapid abandonment, first from the remote to the closest areas of the slope, favoured a rapid development of the successional land.

In Fobello the successional dynamics are controlled by the residual pressure of grazing livestock, which selects the recolonizing tree species and determines the spatial distribution of the regeneration cores. The set of fields, scattered land lots of meadows of big size, determined the occurrence of wide patches of grassland and beech forest (high stands and coppices).

6 Conclusion

The core of the research was an attempt to understand the hierarchy of land use factors influencing landscape patterns at different levels. Thanks to the hierarchical structure of the LUS analysis, it was possible to scale up and down through organization levels towards a better understanding of processes and patterns in space and time. The study of the dynamics of the farming systems through the agricultural inquires and the spatial descriptions of land provided by the cadastres (1863, 1956) bridged the gap between the vegetation dynamics, analysed at the field scale, and the LUS approach. It also bridged a temporal gap, showing LUS structures as an inheritance, determining the actual relationships (topological and chronological relationships) within the landscape unit. Linking observation on landscape and land cover units to farm types, to produce a land cover/use inventory analysis, meant working at different scales with a degree of detail specific to each scale.

Differences between the two valleys were observed at the LUS level (practices and produce), particularly in management – ownership, field size and field combination.

Land use produces landscape in that it produces patterns of land cover: patches, field and associated boundaries, with a certain distribution in space. The similarity of physiographical elements in the Mastallone Basin allowed consideration of historical land use as a key factor in understanding the differences of present vegetation types, secondary succession patterns and land unit arrangements in landscape. The successional dynamics on abandoned pastures and fields were found to be significantly related to original patch size, arrangement and, thus, LUS.

The changes in the last 50 years must thus be viewed in historical perspective. The view that land use and landscape patterns are strongly determined by the physical environment is too simplistic. In the alpine 'constrained' environment of a small basin, and within the same productive frame, *Alpwirtschaft*, local history, settlement structures, choices of farm functioning, productive systems and practices at the field level are major driving factors of landscape organization.

References

Andel, S. *et al.* (1981) Land utilization types for forestry. In: Laban, P. (ed.) *Proceedings of the Workshop on Land Evaluation for Forestry.* ILRI Publication No. 28. Wageningen, pp. 203–205.

Anon. (1976) *A Framework for Land Evaluation.* FAO Soils Bulletin No. 32/ILRI Publication No. 22. Wageningen.

Beek, K.J. (1974) The concept of land utilization types. *FAO Soils Bulletin* 22, 103–120.

Beek, K.J. (1978) *Land Evaluation for Agricultural Development.* ILRI Publication No. 23. Wageningen.

Braak, C.J.F. ter (1987) Ordination. In: Jongman, R.H.G., ter Braak, C.J.F. and van Tongeren, O.F.R. (eds) *Data Analysis in Community and Landscape Ecology.* Pudoc, Wageningen, pp. 91–173.

Braun-Blanquet, J. (1932) *Plant Sociology: the Study of Plant Communities* [translated and edited by G.D. Fuller and H.C. Conard]. McGraw-Hill, New York.

Dansereau, P. (1977) *Ecological Grading and Classification of Land Occupation and Land Use Mosaics.* Geographical Paper Lands Dir. Fish and Env. No. 58. Ottawa, Canada.

Ellenberg, H. (1963) *Vegetation Mitteleuropas mit den Alpen.* Stuttgart.

Ellenberg, H. (1979) *Zeigerwerte der Gefasspflanzen Mitteleuropas.* 2. Aufl. Scripta Geobot. Gottingen No. 9.

Forman, R.T.T. (1982) Interaction among landscape elements: a core of landscape ecology. In: Tjallingi, S.P. and de Veer, A.A. (eds) *Perspectives in Landscape Ecology.* Proceedings of the International Congress of the Netherlands Society of Landscape Ecology, Veldhoven, 6–11 April 1981. Pudoc, Wageningen, pp. 35–48.

Forman, R.T.T. and Godron, M. (1986) *Landscape Ecology.* Wiley, New York.

Jongman, R.H.G., ter Braak, C.J.F. and van Tongeren, O.F.R. (1987) *Data Analysis in Community and Landscape Ecology.* Pudoc, Wageningen.

Kwakernaak, C. (1982) *Landscape Ecology of a Prealpine Area.* Publ. Fys. Geogr. En Bodemk. Lb. Univ. Amsterdam No. 33.

Landolt, E. (1977) *Okologische Zeigerwerte zur Schweizer Flora.* Veroff. Geobot. Inst. ETH, Stiftung Rubel, Zurich No. 64.

Mueller-Dombois, D. and Ellenberg, H. (1974) *Aims and Methods of Vegetation Ecology.* Wiley, New York.

Pielou, E.C. (1984) *The Interpretation of Ecological Data.* Wiley, New York.

Salbitano, F. (1993) Ecological effects on afforestation-studies in the ecology and history of afforestation in Western Europe. In: Watkins, C. (ed.) *Ecological Effects of Afforestation.* University of Nottingham, Nottingham.

Stortelder, A.H.F. and Vos, W. (1983) *Field Guide for the Description of Terrestrial Landscape Ecological Sample Unit.*

van der Maarel, E. (1975) Man-made natural ecosystems in environmental management and planning. In: van Dobben, W.H. and Lowe-McConnel, R.H. (eds) *Unifying Concepts in Ecology.* Junk, The Hague and Pudoc, Wageningen, pp. 263–274.

Westhoff, V. (1983) Man's attitude towards vegetation. In: Holzner, W., Werger, M.J.A. and Ikusima, I. (eds) *Man's Impact on Vegetation.* Junk, The Hague.

Westhoff, V. and van der Maarel, E. (1973) The Braun-Blanquet approach to phyto-sociology. In: Whittaker, R.H. (ed.) *Ordination and Classification of Vegetation.* Handbook of Vegetation Science No. 5. Junk, The Hague, pp. 617–726.

Zonneveld, I.S. (1979) *Land Evaluation and Land (scape) Science*, 2nd edn. ITC Textbook of Photo Interpretation VII.4. ITC, Enschede.

Zonneveld, I.S. (1982) Land(scape) ecology, a science or a state of mind. In: Tjalling, S.P. and de Veer, A.A. (eds) *Perspectives in Landscape Ecology.* Proceedings of the International Congress of the Netherlands Society of Landscape Ecology, Veldhoven, 6–11 April 1981. Pudoc, Wageningen, pp. 9–15.

Temporal and Spatial Changes in a Boreal Forest Landscape: GIS Applications

13

A.-L. Axelsson

Department of Forest Vegetation Ecology, Swedish University of Agricultural Sciences, S-90183 Umeå, Sweden

Historical documents, such as official land surveys and forest inventories, were combined in a geographic information system (GIS) to analyse changes in the forest structure in a large forest landscape in boreal Sweden. We identified forest types with natural structures of great importance for biodiversity (stands with old trees, stands dominated by deciduous trees or stands with a natural multilayered structure) from recent forest inventories as well as old inventories from the end of the 19th century. The development of these key ecological stands were followed over a time scale of more than 100 years.

In the late 19th century, multilayered forest covered most of the forested area in Lycksele parish. Trees older than 150 years were present in 80% of the forested area. The forest was dominated by Scots pine and Norway spruce and the layered structure was created by large recurrent forest fires. Deciduous trees such as birch and aspen occurred naturally in successions after forest fires, as dominant species or mixed with Scots pine and Norway spruce. Due to fire suppression and intensive forest management, that both started at the beginning of the 20th century, the multilayered forest was stepwise converted to even-aged, single-species stands. Herbicide treatment of deciduous trees started in 1950s and led to a major impact at landscape level.

A model on how to incorporate a historical dimension into already existing management plans is presented. The possibility of using historical data to understand the forest dynamics at a landscape level is discussed.

1 Introduction

Northern Sweden is a scarcely populated region dominated by forests; only a few per cent of the land is cultivated farmland. Mires and wetlands cover around 30% of the total area. The forests are dominated by two coniferous species, Scots pine (*Pinus sylvestris*) and Norway spruce (*Picea abies*). There are two major deciduous tree species, birch (*Betula* sp.) and aspen (*Populus tremula*). They occur naturally in successions after forest fires, as dominant species or mixed with Scots pine and Norway spruce.

The population density in this area was below one person km^{-2} until recent times. Early agricultural colonization from the 17th century onwards affected the forest indirectly in different ways; the forest was grazed by domestic animals and used for tar and potash production. The initial logging started in the middle of the 19th century and was a high-grading or selective logging aimed at the largest pines. At the beginning of the 20th century sustained yield management was introduced. Since 1950s clearcutting has been the prevailing logging method.

The forest of Lycksele parish, which Carl Linnaeus (famous as Carl von Linné) travelled through on one of his earliest botanical expeditions, has very little in common with today's intensively managed forest landscape. When young Linneaus visited the area in 1732 he stated that: 'The large forests are desolate and wasteful because no one needs the timber that falls down and decays' (Linneaus, 1732). Today this landscape is characterized by large clearcut areas and there are almost no old or dead trees in the forest. Linneaus would probably be quite surprised if he could repeat his journey now, more than 250 years after his first visit to the area.

The aim of this chapter is to give some examples of temporal and spatial changes at different scales in one boreal forest and present a conceptual model on how to incorporate historical knowledge into modern Swedish forest management.

2 The natural landscape

Fire was the most important natural process in these forests. Most species are well adapted to fire and there are insects and plants that are totally dependent on fire for their reproduction. When the early loggers arrived in the middle of the 19th century there were still large fires burning in this landscape. Earlier studies show that since the 16th century, on average 1% of the forest land was burnt annually by wildfires (Zackrisson, 1977). This gives in average 100 years between the fires. Some of the fires were started by humans to improve the possibilities for grazing in the forests. The reindeer-herding saamis did not use fire as a management tool on a larger scale because lichens, which the reindeer eat during winter, are destroyed by fire. When agricultural colonization started in the 17th century there was a general shift from large natural fires, sometimes covering up to 50,000 ha, to smaller but more frequent fires (Niklasson and Granström, 2000).

Most of the forested area was dominated early on by uneven-aged Scots pine forests with trees of different ages that created a layered structure (Östlund *et al.*, 1997; Axelsson and Östlund, 2000). There were 500-year-old pines standing close together with young and middle-aged trees. Large recurrent forest fires thinned these stands naturally and left pines that seeded the newly burnt areas. After the fire there were good conditions for regeneration and the young trees often grew fast. On dry sites Scots pine regeneration dominated while Norway spruce and birch regenerated on wetter sites. The next fire killed some of the younger trees and left some to seed the area. Most of the old trees survived. Scots pine is well adapted to fire and survives due to a thick layer of bark; Norway spruce is most often killed by fire.

One large ecological change in the boreal forest was the exclusion of natural fires. At the end of the 19th century many roads were constructed and this facilitated an active fire suppression. By the 1920s almost all wildfires were controlled before affecting larger areas (Fig. 13.1). If fires are excluded from this naturally pine-dominated ecosystem, spruce will come to dominate these forests in the long run. To stop what we often call 'sprucification' some national forest reserves in Sweden are today burned after more than 100 years of active fire suppression. Fire is also used for site preparation after clearcutting.

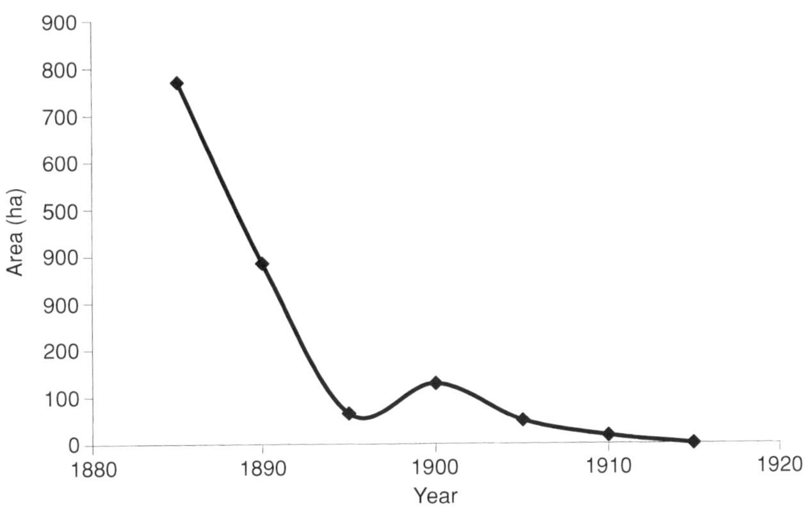

Fig. 13.1. Changes in area burned by wildfire, 1885 to 1915, in Lycksele parish.

3 The managed landscape

So far I have discussed elements of a natural forest landscape that today only exists in small forest reserves and a few national parks. Today most part of the landscape is dominated by large clearcuts and young even-aged forests that form a characteristic landscape pattern. An extensive road system has been built to facilitate timber transportation. Most of the wet sites have been drained. An exotic coniferous species, lodgepole pine (*Pinus contorta*) that is native to North America, has been planted on large areas. Until the beginning of the 1990s, deciduous trees were systematically removed in thinning operations and by using herbicides. It is clear that large overwhelming changes have occurred at landscape level in northern Sweden.

4 Studying the landscape history

We have used historical sources to reconstruct the landscape before large-scale logging operations began. In our study area in northern Sweden the overall ownership pattern has been very stable for over 100 years and there are early official land surveys covering the whole area. In Fig. 13.2, the grey

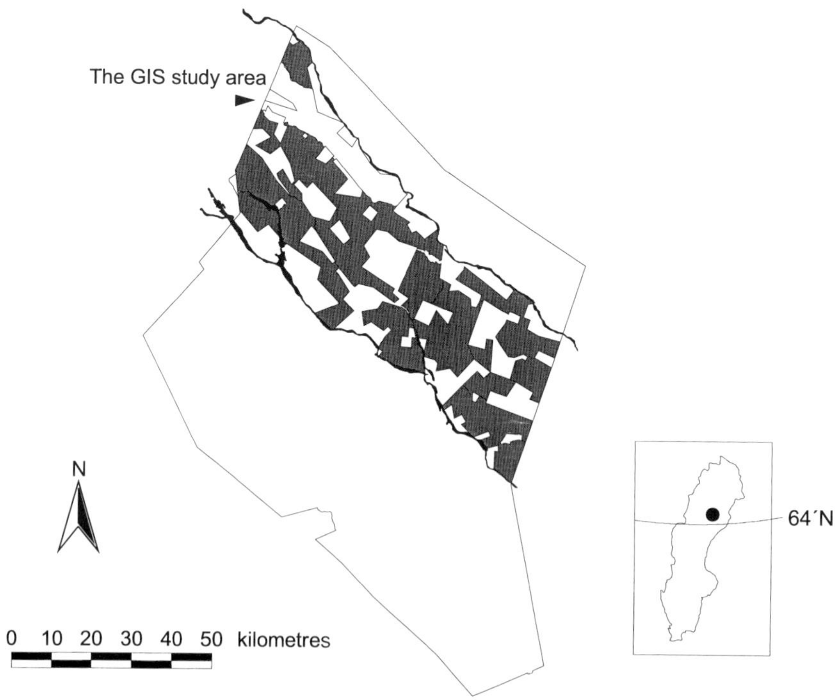

Fig. 13.2. Lycksele parish and the study area chosen for a detailed GIS analysis.

shaded areas represent state forest land and the white areas in between are private land holdings. On state forest land, forest surveys have been performed on a regular basis from the end of the 19th century until today. We have chosen one state forest district for a more detailed geographical analysis. This area covers approximately 170 km^2 (Fig. 13.2).

Old forest survey maps with accompanying descriptions were used in this study. From these surveys it is possible to get detailed information on stand level, concerning species composition, standing volume and age structure. These maps were digitized and combined with a modern digital forest map in a large GIS. Data on natural structures like hydrology, soil and altitude were also included in the GIS. Our earliest data come from a general land survey from 1872 when the areas marked with light grey were recorded as recently burned. By comparing this map to the forest survey from 1914 it is obvious that deciduous trees occurred mainly on two areas that had burned around 50 years earlier (Fig. 13.3). The other deciduous patches refer to fires that probably burned in the 1880s or 1890s after the land surveyors visited the area.

A natural fire does not burn with the same intensity everywhere. Some wildfires, or some parts of the wildfires, burn as low-intensity ground fires that do not kill any trees. After these fires there is no major regeneration. Other more intense fires burned after the first selective logging operations, where logging residuals provided the necessary fuel.

If we compare the situation in 1914 with the survey from 1997 there are large changes. In 1914 about 85% of the forest was dominated by older trees,

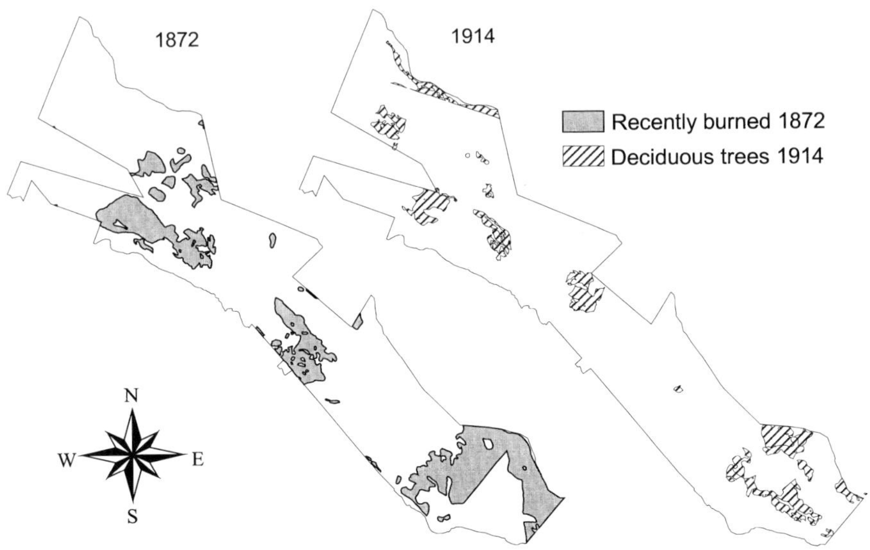

Fig. 13.3. The area burned in 1872 and the area covered with deciduous trees in 1914 in the GIS study area.

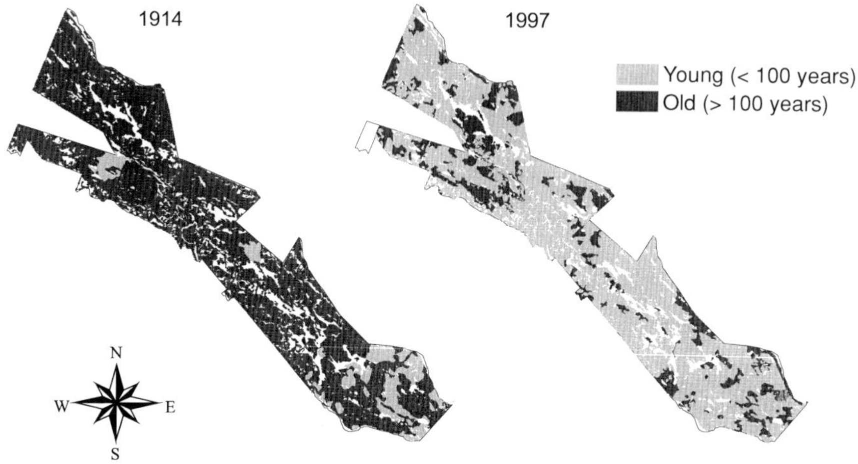

Fig. 13.4. The area dominated by trees older than 100 years in 1914 and 1997 in the GIS study area.

here indicated with dark grey (Fig. 13.4). In 1997, only 20% of the area was dominated by older trees. The natural landscape structure that was created by large forest fires was replaced by a pattern created by clearcuts and forest plantations.

5 A conceptual model for historical GIS

Finally I want to present a conceptual model on how to combine data from different sources into a historical GIS (Fig. 13.5). From the 19th century maps we have information on fire patterns, from the 20th century stand data we have information on species composition, such as sample deciduous forest patches, and from satellite images or the most recent forest survey we can get data on today's forest composition and age structures. Underlying natural conditions can also be integrated to form a complete historical GIS. To be able to compare different historical sources created for different purposes, a generalization of the basic data is needed. Finally, a careful and critical analysis of the historical sources is needed at all stages.

Today, biodiversity issues are on the agenda worldwide. These questions have become a large challenge for forest managers. I believe that retrospective analysis at the landscape level can play a significant role in setting the goals for future forest management where restoration of natural landscape elements is one important aim. By using GIS technology it is easy to incorporate detailed historical data into already existing digital landscape plans. This will create a deeper ecological understanding for important natural processes and

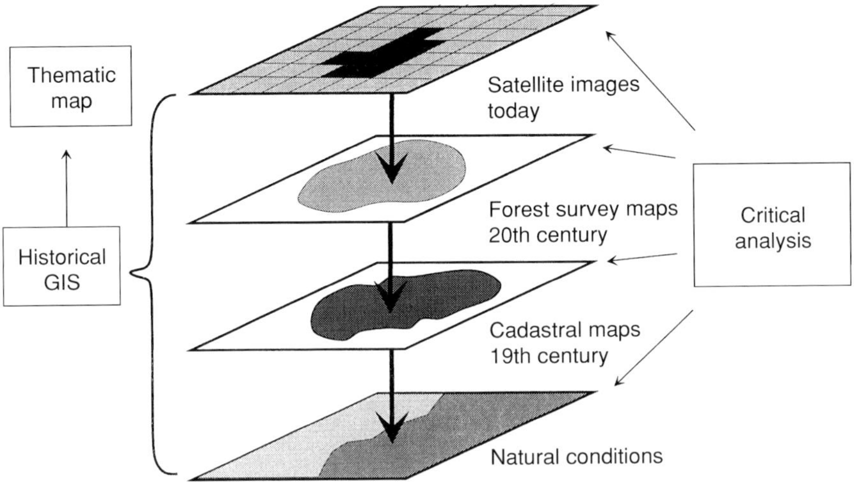

Fig. 13.5. A conceptual model on how to incorporate historical sources and modern forest management plans into a historical GIS.

hopefully guide the forest managers in their efforts to mimic natural disturbance patterns.

References

Axelsson, A.-L. and Östlund, L. (2000) Retrospective gap analysis in a Swedish boreal forest landscape using historical data. *Forest Ecology and Management* (in press).

Linneaus, C. (1732) *Carl Linneaus, Lapplandsresa år 1732* [Caroli Linneai. Iter Lapponicum 1732]. Wahlström och Widstrand, Stockholm 1975.

Niklasson, M. and Granström, A. (2000) Numbers and size of fires: long term spatially explicit fire history in a Swedish boreal landscape. *Ecology* (in press).

Östlund, L., Zackrisson, O. and Axelsson, A.-L. (1997) The history and transformation of a Scandinavian boreal forest landscape since the nineteenth century. *Canadian Journal of Forest Research* 27, 1198–1206.

Zackrisson, O. (1977) Influence of forest fires on the north Swedish boreal forest. *Oikos* 29, 22–32.

Comparing Damages: Italian and American Concepts of Restoration

14

M. Hall

Institute for Environmental Studies, University of Wisconsin, Madison, WI 53706, USA; present address: Swiss Federal Institute of Snow and Landscape Research (WSL), 8903 Birmenodort, Switzerland.

The endeavour of environmental (or ecological) restoration is much older than a few years or decades, as people have long been converting damaged land into ideal former states. One can also detect national styles of restoration. Over the last 100 years in the United States the goal of restoration has often been to *re-wild* the land by excluding detrimental human elements, whereas the usual goal in Italy has been to *re-garden* the land by including beneficial human elements. These different national styles are due, in part, to different visions of damage to the land. Whereas Americans typically focused on repairing degraded lands that were damaged by human processes, Italians often focused on improving degenerated lands that were damaged by natural processes. Restoring by repair as opposed to restoring by improvement had important consequences for land management. By drawing on historical examples of restoration in both countries, it will be shown that the restorative techniques and the restored lands depended on different national concepts of degradation. These differing views of degradation continue to have important consequences for current environmental restoration.

1 Introduction

According to at least one source, impoverished and abandoned lands are today's fastest growing class of land globally (Wali, 1992). From deforested jungles, open-pit mines and empty city lots, to fallow fields, overgrazed pastures and eroded watersheds, much of this damaged land must be rehabilitated or reclaimed to be of further use to humans and other earthly inhabitants. This endeavour of environmental restoration takes many forms and many degrees and can last for months or decades, and can involve a few people or whole government divisions. Depending on the project, a series of decisions must be made about how to restore the land. One of the most commonly debated questions concerns the condition to which the land should be restored. Restorationists in North America often set their goal on restoring pre-European settlement conditions when the land – though already inhabited – was supposedly more 'pristine', 'wild', or 'natural' (Covington in Nijhuis, 1999). But Europeans and other Old World inhabitants must seek different criteria for the goals of their restoration projects. Rather than trying to recreate 'presettlement' or 'indigenous' lands, many Europeans hope to bring back the land's previous beauty or integrity. With input largely from Americans and Europeans, the Society for Ecological Restoration declared in 1995 that restoration is 'the process of renewing and maintaining ecosystem health'. But renewing and maintaining healthy states of nature is not a new pursuit or a novel idea.

Restoring nature is a very old activity. For at least the last 200 years, the early literature on land management in the US and abroad is replete with references about ways to restore the land. Early foresters, civil engineers and agricultural theorists often spoke of 'restoring' the woods, the soils or the mountains. The ideas encapsulated by restoration and its synonyms such as reclamation, reconstruction and improvement – together with numerous equivalent foreign terms – begin to show the great variety of ways that land managers have sought to convert damaged land into former ideal states. This chapter presents three models, or views, of restoration that depend on the inputs and outputs of the restorative process. To illustrate these models, I will offer a few examples of restoration across time and place, using the contrasts of Italian and American land management to help clarify these models. My other goal is to show what restoration might reveal about our assumptions about degradation, and how such assumptions change the way we manage lands.

I have already suggested that there seem to be different ideal outputs to restoration in Europe and North America. Italians, for instance, traditionally hope to restore land to ideal *domesticated* conditions that require human presence, while Americans hope to restore land to ideal *primitive* conditions that require human absence. One can say that Italians have long sought to restore gardened land, while Americans have sought to restore wildland. In 1918, the noted Italian civil engineer, Federico Sacco, advocated reforesting his country's barren hillsides in a process much like gardening: 'Italy, called the Garden of Europe, but now unfortunately reduced in part to a garden without trees, includes a half million hectares of land that lie uncultivated, sterile, useless, even dangerous. . . .[This land] must be returned to the forest economy in order to utilize its hydraulic and agricultural potential' (Sacco, 1918). By contrast, the American plant ecologist and range manager, Arthur Sampson, sought ways to convert grazing lands into a primordial climax state in a process that might be called naturalizing: 'on badly eroded or denuded range, we desire to re-establish the climax species . . .' (Sampson, 1917). This national characterization of Italian and American restoration as either gardening or naturalizing, surfaces repeatedly in the writings of land managers from each country since the early 20th century.

To further illustrate these national differences, Americans typically favoured native species in their wildlands, whereas Italians typically favoured exotic species in their gardened lands. A leading Italian forester, Aldo Pavari, often praised the benefits of exotic forest species, as when he wrote a treatise on the subject in 1915. But the US Department of Agriculture declared in its Yearbook of 1915 that the best remedy to halt erosion in grazing lands is 'to restore the native vegetation of the pasture' (Jardine, 1915). Even landscape architects and park planners can be categorized according to these national styles of gardening or naturalizing: in their gardens and city parks, Italians have become famous for developing a 'formal' style that reflects human order and control, whereas Americans have become famous for advancing a 'naturalistic' style that appears more like that which can be found in nature. This national difference of favouring domesticated or wild nature is also reflected in the management of national parks – Abruzzo National Park incorporates inhabitants within its borders, whereas Yellowstone National Park excludes permanent inhabitants. Gardening on the one hand, and naturalizing on the other, are broad though useful ways to describe the main differences between Italian and American land restoration.

2 Damage

But if restoration produces outputs, it also requires inputs. Not only did Italian and American land managers confront the question, 'Restore to what?', they also confronted the question, 'What to restore?' In other words, land managers in each country first had to identify damaged land – or damage in the land – before they attempted to correct it. For example, deforested watersheds often merited restoration in both countries because such areas frequently flooded, causing loss of life and property, and so represented damaged land in need of a remedy like reforestation. As early as 1816, Francesco Mengotti had taught that floods could be prevented by revegetating barren watersheds. Yet it is easy to see why not everyone considered deforestation to be damage. Sheep and cow herders, after all, benefited from pastures created by felling trees. Like ideal land, damaged land was open to interpretation. In their restoration projects, Italians and Americans often targeted different types of damaged lands because they envisioned different ways of doing damage to these lands.

In reviewing early descriptions of various restorative projects, or when walking through restored lands in each country, one can conclude that Italians often saw damage as *degeneration*, but that Americans saw damage as *degradation* – degeneration resulted from spontaneous non-human forces; degradation resulted from human activities. Consider mountain floods. When the Alps or Apennines flooded a century ago, most Italians envisioned the mountain to be degenerating under the forces of wind and rain: people were to blame only because they had failed to plant and maintain stabilizing forests. But when the Rocky mountains or Appalachians flooded, Americans envisioned the mountain to be degraded by human activities: people were to blame because they had cut down the forests. Since the late 19th century, Italians spoke of decay and deterioration, Americans spoke of mistreatment and misuse. From the American perspective, degradation resulted from human *action*, but from the Italian perspective, degeneration resulted from human *inaction*. In both countries, citizens called for reforesting and rebuilding flood-prone watersheds; the US Forest Service and the Italian Corpo Forestale planted trees and grasses in order to restore stability to their mountains. But while Italians sought to restore human order to managed forests that had fallen into decay, Americans sought to restore natural order to primordial forests that had been exploited. By seeing degeneration, Italians blamed nature more than culture; by seeing degradation, Americans blamed culture more than nature. Over the last two centuries, just as Americans have been faster than Italians to see ideal land as human-free, Americans have also been faster to see damaged land as human-caused. I believe that these notions about ideal and damaged land stemmed from national myths of the United States as a tamed wilderness, and of Italy as a manicured garden. Just as Americans have learned to see untouched wildland as one of their country's greatest assets, Italians have learned to see rich cultural landscapes as one of their country's greatest wonders. Americans still feel that their wildlands are threatened by people; Italians still feel that their gardens are threatened by the absence of gardeners. Both countries have extensive humanized and natural landscapes, but Italians and Americans point to different heroes and different villains when they see changes in those landscapes.

As the following images may help illustrate, Italians traditionally counter damage by *improving* degenerated land, while Americans counter damage by *repairing* degraded land. Gardening and naturalizing (which depend on outputs) and improvement and repair (which depend on inputs) are the best ways to contrast the two national styles of restoration. At the turn of the 20th century, one might say that Italians restored by *gardening degenerated* land while Americans were beginning to restore by *naturalizing degraded* land – the intermediate view of restoration, *gardening degraded* land has been favoured at different times by land managers in both countries.

3 Three views of restoration

These different views of restoration might be represented graphically so as to clarify two very different types of damage. Figure 14.3 shows what can be called, 'gardening the degenerated' and Fig. 14.2 'naturalizing the degraded'. After staring at these two models, one can envision at least one other view of restoration that is intermediate between the other two: 'gardening the degraded' (Fig. 14.2) is a hybrid view of restoration in which land managers not only blame human abilities to damage land but praise human abilities to create ideal land.

Finally, if one reviews restoration projects during the last 50 years, Italians have increasingly seen culture (rather than nature) as the main source of land damage, at the same time that Americans have increasingly seen nature (rather than culture) as the main source of ideal land. I believe that over this period Italians were changing their view of restoration from that shown in Fig. 14.1 to Fig. 14.2, while Americans were changing their own view of restoration from Fig. 14.2 to Fig. 14.3. Stated differently, Italians have recently been getting used to the idea that culture does the most damage to land and Americans have been getting used to the idea that nature provides the most benefit to land. Today, it is safe to say that while Italians still dread natural damage to their garden, Americans still dread cultural damage to their wildland. Italians rarely think of restoration in terms of Fig. 14.3; Americans rarely think of it in terms of Fig. 14.1.

4 So what?

I have not spent much of my time here focusing on how these conceptual changes took place. The reasons *why* Italians and Americans have shifted their views about the restorative process vary from region to region and project to project. It may be the case that restorationists in both countries have simply been finding more ways to restore, rather than substituting one view for another. Today, I think we see restorationists who garden or naturalize, repair or improve, according to each of the three models. But if one looks through early management projects that were labelled as restoration – whether these involve reforestation, soil conservation, sand dune stabilization, desert or marsh reclamation – Italians usually saw nature as the problem, while Americans saw people as the problem. Americans have been more willing than Italians to see ideal land as human-free, just as Americans have been

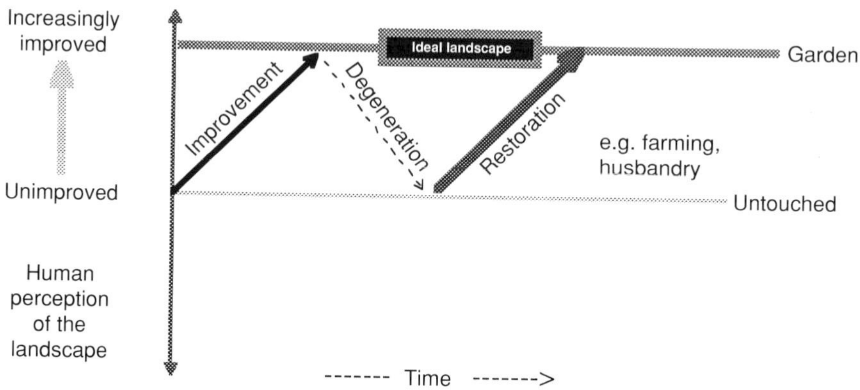

Fig. 14.1. 'Gardening the degenerated'.

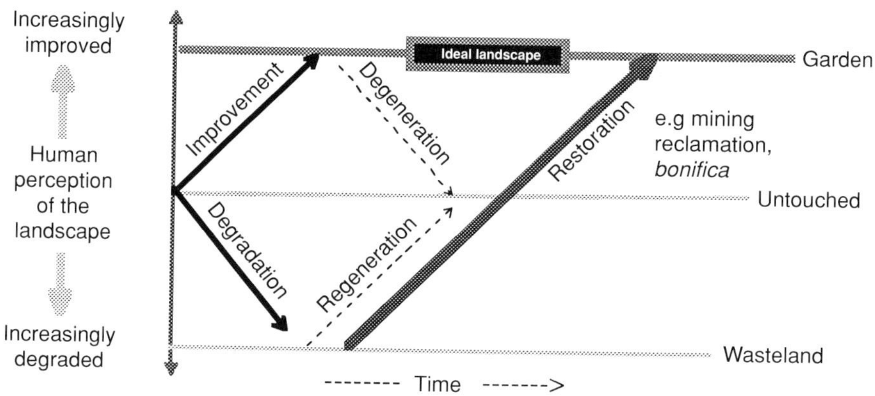

Fig. 14.2. 'Gardening the degraded'.

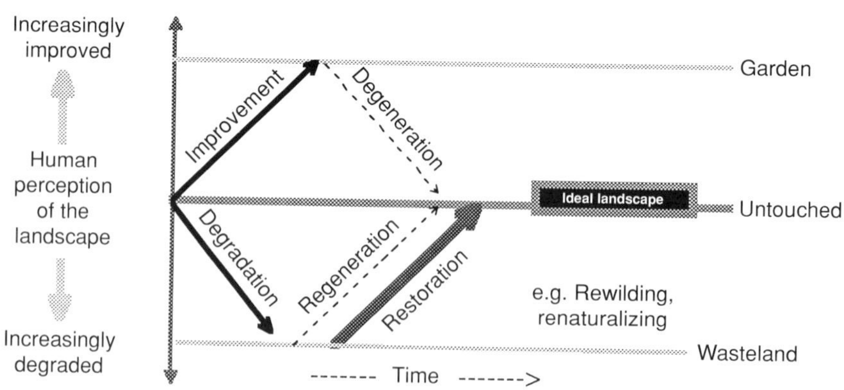

Fig. 14.3. 'Naturalizing the degraded'.

more willing to see damaged land as human-caused. Much as Marjorie Hope Nicolson traced western attitudes about mountains from gloom in the 17th century to glory 200 years later, I believe that one could trace western attitudes about land damage from degeneration to degradation – from blaming nature to blaming culture (Nicholson, 1959).

These two different ways of envisioning land damage have implications for land management. American land managers are used to damning people, whereas Italian land managers are used to damning nature – when Italians do blame people for the poor condition of their land, they blame them indirectly and secondarily. Most Italians consider human inaction, not human action, as the major threat to their lands. Although many Italian and American landscapes may look the same, and may be equally damaged, different national perceptions about sources of damage mean that restorative strategies can vary widely.

Land managers in both countries may therefore improve their craft by reconsidering environmental damage. Natural forest fires may occasionally benefit Italian forests, just as human fire suppression may occasionally benefit Yellowstone's forests. Stated differently, Americans might learn from Italians that natural damage is not necessarily benign (as when Yellowstone burns) or that human damage is not necessarily evil (as when loggers harvest trees). Likewise, Italians might learn from Americans that natural damage is not always evil (as when forest fires are allowed to burn) or that human damage is not always benign (as when agriculture diminishes biodiversity). I do not mean to advocate a style of restoration whereby any one land use practice can be justified; rather I hope that we can identify the source of damage so that we know how best to approach restoration efforts.

If environmental restoration reveals that Americans praise nature, it also reveals that they blame culture. There seems little doubt that the American love of wilderness is linked to the American fear of degradation. Rather than *re-wilding* the land, American restorationists might better see themselves *re-gardening* the land. Or if Americans could possibly see their wildlands as gardens, then they might become better at restoring these lands. Americans cannot understand how Europeans *re-nature* their land without reference to a wild landscape, and Europeans cannot understand how Americans *restore* their land without reference to a cultural landscape; it may simply be that both sides see different sources of land damage.

References

Covington, W. quoted in Nijhuis, M. (1999) Flagstaff searches for its forests' future. *High Country News* 31, 8.

Jardine, J.T. (1915) Improvement and management of native pastures in the west. *Yearbook of the United States Department of Agriculture*. GPO,Washington, DC, p. 310.

Mengotti, F. (1816) *Idraulica Fisica e Sperimentale; ossia saggio sull'aque correnti*, Vol. I. Andreola, Venezia, pp. 281–310.

Nicolson, M.H. (1959) *Mountain Gloom and Mountain Glory: the Development of the Aesthetics of the Infinite.* Cornell University Press, Ithaca (reprinted University of Washington Press, Seattle, 1997).

Pavari, A. (1914–1915) Studio preliminare sulla coltura di specie forestali esotiche in Italia. *Annali del R. Ist. Superiore Forestale Nazionale* I, 159–380.

Sacco, F. (1918) *La Sistemazione Idrico-Forestale dei Bacini Montani.* Nistri, Pisa, pp. 78–90.

Sampson, A.W. (1917) Succession as a factor in range management. *Journal of Forestry* 15.

Wali, M.K. (1992) *Ecosystem Rehabilitation: Preamble to Sustainable Development.* SPB Academic Publishing, The Hague.

Searching for Common Ground: Reconstructing Landscape History in East Africa's Eastern Arc Mountains

15

C.A. Conte

Utah State University, Department of History, Logan, UT 84322, USA

This chapter argues for the extension of forest history studies from the microstudy to regional comparative examinations across regions with continuities in natural and human history. It contends, further, that effective resource conservation requires a complex understanding of change along temporal and spatial scales that include long-term histories of land use and transformation in physical and biological landscapes. After a brief review of the relevant literature on environmental change in Africa, the chapter makes the case for studying the continuities and differences in the environmental history in Tanzania and Kenya's Eastern Arc mountains, arguing for the employment of a bioregional approach and a set of flexible concepts that would allow a discussion across disciplinary boundaries. The final section also examines relevant types of evidence then argues for the practical utility of this type of historical study to forest conservation policy.

1 Introduction

Nature has endowed East Africa's mountain ecosystems with more moisture and cooler temperatures than their surrounding plains. These physical endowments have fostered millions of years of montane forest evolution that has produced pockets of rich biological diversity (Lovett, 1993; Wasser and Lovett, 1993). Over the last several millennia, mountain resources have also helped sustain agrarian peoples' ability to mitigate the capriciousness of periodic droughts characteristic of East Africa's climate. However, over the last century, research scientists and bureaucrats associated with both the colonial and independence governments have questioned the sustainability of these mountain-based indigenous agricultural systems and have tried, often unsuccessfully, to transform the interaction between farming communities and what the experts argue is an increasingly degraded environment. Though the assumptions behind the degradation discourse have been the subject of increasingly sophisticated scholarly analyses, soil, water and forest conservation projects based on ahistorical readings of ecological change continue to hold sway in the development community. This chapter contends that effective resource conservation requires a complex understanding of the long-term history of land use and landscape transformation. Furthermore, it points out that useful historical interpretations of resource use need to be both undertaken on a regional scale and rooted in the landscape (e.g. McCann, 1995).

After a brief review of the relevant literature on environmental change in Africa, the chapter makes the case for studying the continuities and differences in the environmental history in Tanzania and Kenya's Eastern Arc mountains, arguing for the employment of a bioregional approach and a set of flexible concepts. The final section examines relevant types of evidence then argues for the practical utility of this type of historical study to forest conservation policy.

2 Recent studies surrounding ecological wisdom and landscape history

Understanding ecological transformations on forested, and formerly forested, landscapes requires an accounting of varying cultural interpretations of land use history and their placement into the context of a changing physical and biological environment. However, doing so presents complex interpretive problems that challenge the training of most social scientists, the scholarly pronouncements on the need for interdisciplinary research notwithstanding. A brief review of the Africanist literature none the less reveals an increasingly sophisticated scholarship that is moving in the right direction.

Recent studies in geography and social anthropology, for example, have examined the validity of heretofore widely accepted orthodoxies regarding deforestation, soil erosion, overgrazing and desertification in Africa. In their introduction to a collection of essays called *The Lie of the Land* (1996), editors Melissa Leach and Robin Mearns suggest that the western tradition of post-enlightenment scientific investigation has created a set of ideas about ecological change which filtered into colonial interpretations as an unquestioned 'received wisdom'. They argue that these notions, which often form the foundations of resource conservation policy, have led to widely accepted but erroneous assumptions about the causes, or even the existence, of ecological stress and degradation in non-western contexts. James Fairhead and Melissa Leach's book-length investigation, *Misreading the Landscape* (Fairhead and Leach, 1996), employs these ideas in reconstructing change on an agricultural landscape in Guinea's ecological transition zone where forest meets savanna. Using aerial and satellite photos, along with a careful on-the-ground assessment of local land use practices, they argue quite convincingly that contrary to the presiding historical interpretations of deforestation, local practice had actually fostered an increase in forest cover.

The study begs some important questions however. *Misreading the Landscape* is an analytically sophisticated, yet local, study of changing ecological conditions. One is left asking, therefore, how Fairhead and Leach's results compare with other similar environmental situations across Africa. If the lack of broader analyses ultimately points out our ignorance of regional patterns of vegetation change, microstudies clearly illuminate just how generalized views condition landscape readings. A natural resource policy more sensitive to historical trends and useful regionally might well benefit from future analyses that build local complexity into regional studies of environmental change. Bioregions, i.e. regions that show continuities in human and natural history, lend themselves particularly well to such an historical analysis. In addition to scales of analysis, Fairhead and Leach's reading of the colonial scientific literature also raises important issues regarding historical evidence. Should scientific explanation authored under colonial regimes simply be jettisoned on ideological grounds? Historians, for example, might ask if the generalized views of high-ranking French colonial foresters regarding West Africa's deforestation went unchallenged within either the forest service or the wider scientific community. If not, what were the institutional barriers to dissent? If so, what did other analyses reveal?

This chapter submits that the value of microstudies increases as they are incorporated into a broader regional and temporal framework. In a recent collection of essays called *Historical Ecology* (Crumley, 1994), several authors address time scale as they seek to uncover the long-term nature and scope of ecological change and degradation. Like the above-mentioned 'received wisdom' studies, the *Historical Ecology* essays attempt to reconstruct social and ecological change across landscapes and are grounded firmly in anthropological research. However, the book's editor, Carol Crumley, explains that historical ecology, as an analytical approach to the study of the functional and structural properties of ecosystems, seeks purposefully to develop a language that bridges the social and natural sciences (Crumley, 1994). Environmental historians of eastern Africa have begun to incorporate into their analyses a longer time frame, but archaeological scholarship has, to this point, offered some of the best examinations of pre-20th century landscapes (for example, Sutton, 1984; Waller, 1990; Schoenbrun, 1993; Schmidt, 1994). Peter Schmidt (1994), for example, has examined the interaction of nature and culture over the long term in the vicinity of Lake Victoria. Schmidt's research stresses the relationship between natural resource management, especially forest endowments, and 2000 years of demographic and technological change. David Schoenbrun's (1993) historical study of the same general region uses with great methodological innovation historical linguistics and pollen analyses to reconstruct pre-15th century land use and culture. Taken together, the Lake Victoria studies offer a deeply layered interpretation of regional patterns of environmental and cultural change. Schmidt (1994) and Schoenbrun (1993) depart from Fairhead and Leach's (1996) scenario for Guinea and argue for a much more dramatic transformation in which large-scale deforestation accompanied technological innovation in iron-working technology.

John Sutton, an archaeologist with long experience looking at historical East African landscapes, has analysed long-term ecosystem change along the Rift Valley escarpment, where the florescence and breakdown of intensive agricultural systems is still written on the landscape (Sutton, 1984). Sutton's work is especially relevant to agricultural and landscape history in the nearby Eastern Arc mountains, where agricultural production depended on a similar complex of rain-fed and irrigated gardens, intercropping, and the exploitation of forest soils spanning a mountain elevation gradient. Schmidt, Schoenbrun and Sutton's arguments for the continuities in husbandry practice across space and time suggest that regional, rather than local production patterns linked East African farming communities.

Placed in a long-term context, these historical interpretations make the case for centuries of adaptations in husbandry in particular environmental contexts that at times reach a point of diminishing returns that results in environmental degradation, necessitating either technological innovation or migration, or perhaps a combination of both. The studies cited above suggest a fruitful approach for recovering ecological history which is broad in regional and temporal scope, but which is sensitive to local variation. Moreover, given the ongoing conflict over land, timber and water in eastern Africa, this history could inform a conservation policy sensitive to the complexities in indigenous patterns of resource conservation and degradation.

3 Using history: recovering landscape change in the Eastern Arc mountains, the argument for unity across a bioregion

East Africa's coastal mountains (referred to here as the Eastern Arc mountains; see Fig. 15.1) present a particularly useful field site for a comparative study of the history of land use adaptations in and around tropical forests. Their study also presents an opportunity to understand long-term change in forest composition and use and to inform resource management in some of East

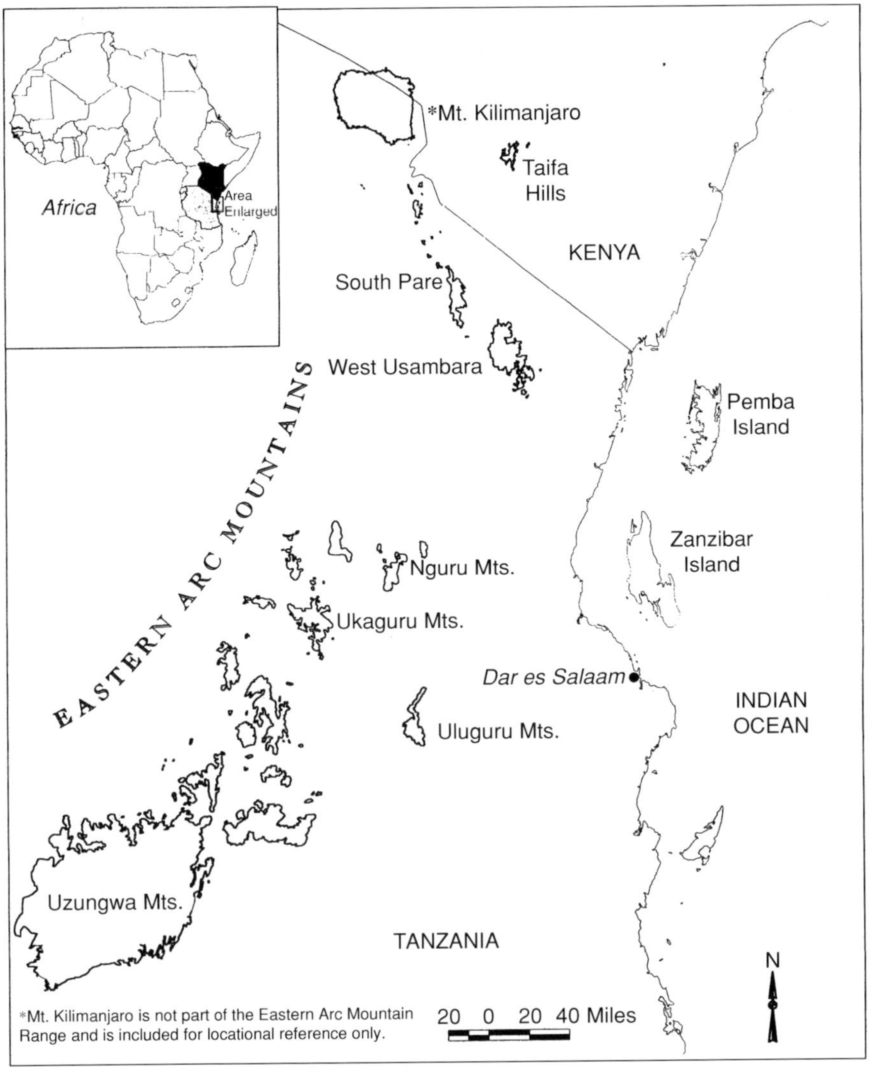

Fig. 15.1. Eastern Arc mountains.

Africa's most densely populated areas. The geologically ancient crystalline formations of the Eastern Arc have carried forest for the past 30 million years. Their biological value stems from geological processes through which they became isolated from forests in Central and West Africa. As the Eastern Arc forests became, in effect, biogeographical islands, plant and animal species evolved along their own pathways creating high rates of endemism.

Moreover, the moist forests on the eastern slopes of the block-faulted mountains continued to condense the moist air made available by the Indian Ocean, whose climate seems to have remained relatively stable through the temperature fluctuations of the Pleistocene.

Complex mountain topography scatters the rains unevenly creating microclimates that ultimately affect plant make-up and associations. In sum, the mosaic of Eastern Arc forest communities correspond closely with aspect and elevations from the low- and medium-altitudes of the mountain bases (300–400 m) to the high mountain plateaus, sometimes reaching 2000 m. Under conditions of heavy annual rainfall, low- and medium-altitude forest can be very tall with emergents to 60 m whereas high-altitude ridge-top forests several miles away may have a closed canopy only 2–3 m tall. Moreover, local rain shadow effects create a variety of well-defined forest types at similar altitudes (Lovett, 1993). These forest conditions have fostered the evolution of hundreds of endemic species. In the Usambara mountains, 25–30% of the *c.* 2000 species are endemic with 16 genera endemic or near endemic.

In terms of human history, the contiguous nature of these forest zones presented agricultural communities with attractive environments in which to produce crops with very different moisture and temperature demands. By the turn of the last millennium, the mountain forests had become a refuge for Iron Age East African farmers in search of new lands. Early in their settlement history, they exploited very specific sites near the massif edges (Schmidt, 1987). These mountain settlements belonged to what archaeologists have referred to as the *Mwitu* tradition of the East African Early Iron Age, which spread from the iron-using communities of the African Great Lakes region beginning about 2000 years ago. In the Usambara mountains the associated *Mwitu* complex seems to have become prevalent sometime during the 2nd century AD (Soper, 1967). Over the course of several centuries, Early Iron Age peoples of the Eastern Arc mountains forged their own but related cultural identities and developed economies dependent on agriculture, iron working and exchanges with neighbours who exploited the plains ecosystems (Vansina, 1995). That the colonization occurred in an environment significantly different from its lacustrine base begs the question of how the spread occurred and specifically what cultural innovations were necessary in adapting to the relatively drier climate and more acidic soils of the Eastern Arc mountains. In any event, the removal of forest cover for settlement and agriculture constituted a dramatic and ongoing ecological process in the Eastern Arc mountains.

The admittedly limited archaeological evidence suggests that *Mwitu* communities in the Eastern Arc originally confined themselves to the wetter montane or sub-montane woodland/forest areas (Collett, 1985; Schmidt, 1987). The distribution of pottery sherds suggests, however, a gradual spread down the hills on to the drier plains. Movements to the more marginal agricultural areas may have been spurred by population growth, a wetter climate, deforestation and the resulting movement of topsoil from hills to plains, crop innovations, or a combination of all four. *Mwitu* peoples clearly understood the advantages and risks of exploiting a variety of ecological situations and situated themselves at the edges of the massifs to take advantage of the possibilities offered by their cultivation at both high and low elevations (Collett, 1985; Schmidt, 1987). In the northwest side of the West Usambara mountains, for example, farmers still cultivate both the alluvial soils of the upland basin along with their gardens 2 hours' walk away on the plains below the escarpment (Conte, 1999). As they know from centuries of experience, less reliable rainfall in the plains heightens their risk, but higher temperatures significantly shorten the growing season allowing for a quick and profitable harvest when rains prove adequate.

The affinities with the lacustrine Iron Age communities, and essential similarities among Early Iron Age settlement patterns and land use practices in the Eastern Arc suggest that settlement patterns represent a maturing of a complex of production systems tailored to new environmental contexts. One finds the artifacts of shared knowledge among Eastern Arc peoples expressed in their related languages, the similarity in their pottery styles, their propensity to farm similar ecological niches and their history of iron working. The argument for interaction among the peoples of the Eastern Arc mountains is further buttressed by migration stories in Usambara, where the Shambaa ethnic group contains various clan divisions which claim historical ties to other nearby Eastern Arc mountain regions (Feierman, 1974, Chapter 2).

Research to date on the history of Eastern African highland forest agriculture suggests that farming systems were highly adaptive. Iron Age agricultural production in the Eastern Arc mountains clearly experienced several important transformations during the course of the early and later Iron Age (post-1000 AD) (Conte, 1999). These transformations would have been expressed in terms of cultigen choice, technological innovation and intensification. Change occurred under currently little understood demographic circumstances, and while they exhibit a strongly regional flavour, resulting environmental change differed, not only from massif to massif, but within each massif (Vansina, 1994/95). Even a preliminary look reveals important differentials of degradation and the more detailed comparative study proposed here will, I believe, provide clues to effective conservation approaches. A comparative environmental history, valuable in its own right as a statement on the human condition, thus becomes essential to regional development policy that seeks to avoid blanket solutions to complex problems. In a broader sense, the reconstruction of northeastern Tanzania's mountain history will be applicable to other cases in the tropics where indigenous and western approaches to resource conservation have come into conflict.

4 Reconstructing forest history in the Eastern Arc: preliminary research

An examination of historical ecology in the Usambara mountains suggests that a regionally broader and more comparative approach to the history of land use and landscapes is possible for the Eastern Arcs. My doctoral research examined changing land use patterns under German and British colonialism in Tanzania's West Usambara mountains. From preliminary interviews in 1991, I developed and continuously refined a set of questions designed to elicit responses regarding transformations in plant and animal husbandry, resource conservation, land tenure, labour organization, rainmaking practice and local politics. I also chose to interview in a variety of ecological zones that cross the elevation gradient from the mountain plateau to the plains below the massif. My analysis of the oral and documentary evidence revealed that processes of environmental change, stress and degradation are extremely complex. Moreover, it showed that indigenous farmers and forest-based herders viewed natural resource endowments in ways strikingly different from their contemporaries in the colonial administration who forced upon them rules and regulations for soil and forest conservation.

During a 1996 research trip to Tanzania, I began to explore possibilities for comparative research in the mountain regions neighbouring West Usambara. On the surface, at least, I found clear differences in the severity of environmental degradation on mountain massifs which had broadly similar geological, climatic and human histories. It became clear to me that a comparative study of mountain ecological history could inform the official resource conservation and development efforts so ubiquitous across the East African highlands. With this in mind, I contacted the leaders of the East Usambara Catchment Forest Project (EUCFP, a cooperative venture between the Finnish and Tanzanian governments) and through visits to several East Usambara villages, I began to assess the feasibility of collecting oral testimony there.

Subsequent fieldwork in East Usambara during autumn 1998, however, revealed just how politically charged environmental issues had become since the May announcement of the Amani Nature Reserve. After several days of interviews, it became clear that the villagers in the East Usambaras were reluctant to speak with us because they were extremely upset about events surrounding the creation of the East Usambara Rain Forest Nature Reserve. Apparently, the EUCFP researchers had surveyed farms and promised farmers monetary compensation for lands removed from cultivation and turned over to the Nature Reserve. Some informants argued that they had not been compensated, even though they believed that the Finnish Government and the European Union, two of the sponsors of the EUCFP, had disbursed funds to the local district authorities. Whatever the reality, rumors in the mountains had fed a visceral sense of ill will among some farmers toward the newly adjudicated Nature Reserve. The fact that my assistant, also associated with the EUCFP, was the local government forester charged with policing this new reserve, complicated matters for us. These associations heightened local suspicions of our motives in many, although not all, of the villages and hamlets we visited.

The difficult circumstances forced us to reconsider my research plan. My local contact agreed to continue with the interviews but to limit his questions to less controversial, but none the less useful topics revolving around historical migration patterns into the East Usambaras. While my associate remained in the mountains, I returned to Dar es Salaam to study the colonial files on conflicts over land use contained in the Tanzania National Archives in the area. This investigation showed that clearly we were not the first observers in the East Usambaras to note the existence of heated land disputes.

The field experience proved extremely valuable in that it reinforced in my mind the absolute necessity of remaining for an extended period of time in a particular area in order to gather oral history, as I had done in the West Usambaras. It also brought home the reality of the volatile nature of land issues in rural East Africa, where there are at present no title deeds to most agricultural lands (historically known as 'Public Lands') and where access and ownership are still locally negotiated.

5 Concepts, evidence and questions[1]

5.1 Landscape

Interdisciplinary collaboration among anthropologists, historians, ecologists and natural scientists is impossible without a set of flexible concepts that inform analyses of landscape transformation (Winterhalder, 1994). This study attempts in a preliminary fashion to map out a meaningful and flexible language of analysis. Spatial concepts are particularly important to reconstructing vegetation patterns and analysing ecological change at scales ranging from soil structures in a farmer's garden to mountain forests. Of primacy in this work is the concept of a landscape, which can be examined as both a cultural and a physical phenomenon. In terms of their physical manifestation, landscapes can be delineated at a variety of scales but none the less contain a pattern of repeating environmental and/or cultural elements. The landscape concept has particularly apt applications in the Usambaras where ecological communities vary greatly across space (with elevation and aspect) and time (according to land use patterns). In their cultural guises, landscapes serve as important parts of historical and ideological reckoning. By employing landscape as a central concept in a comparative perspective, this study

[1] The following discussion of concepts relies heavily on Winterhalder (1994).

problematizes historical generalizations about environmental degradation and rehabilitation in East Africa.

Another useful spatial concept is the patch, an ecologically distinct feature embedded in a landscape. In the forested and formerly forested zones of the Eastern Arc, patches are distributed across a particular landscape as heterogeneous ecological communities and can demonstrate a history of land use patterns nested within a broader landscape. Similarly, the ecotope, or an ecosystem with homogeneous properties (e.g. a section of forest among agricultural fields), can reveal a complex history of conscious land use planning or conservation of specific vegetation communities for ritual purposes among local communities, as the work of Fairhead and Leach, cited earlier, reveals. Spatial concepts are particularly important to explanations of changing vegetation patterns that employ geographic information systems to map and interpret evidence from aerial and satellite photogaphs. They are likewise important because they create a linear framework within which one can integrate qualitative evidence from oral interviews and archival data.

5.2 Historical process, degradation

Leach and Mearns (1996) argue that one weakness of scientific views of African land use has been their assumption of systemic stasis and inflexibility. Historical interpretations, on the other hand, seek out dynamism in land use practice, culture and landscape. None the less, dynamic approaches must recognize associated conceptual problems; the meaning of ecological concepts changes over time. As they have with other concepts, like carrying capacity and succession, environmental historians have regularly invoked environmental degradation without a clear understanding of its continually shifting definition. For example, definitions of degradation associated with the disturbance of heretofore homeostatic ecological communities (as with 'climax communities') are problematic, especially when they consider as axiomatic a link between land degradation and increases in human populations, or between degradation and the growth of capitalism. This study accepts that degradation describes the diminution of a particular place's ability to produce biomass. However, it does so with the understanding that degradation is a process and only one of many possible trends in measures of exploitation and bioproductivity, rather than a departure from a hypothetical baseline of stability (Mortimer, 1998).

By concentrating conceptually on degradation, a study of the Eastern Arcs could account for the historical trajectory of a colonial and post-colonial resource conservation policy that has historically aimed to re-create a perhaps mythical ecological stability, as well as the paradox that policy has often been accompanied by environmental degradation in the diminution of biological diversity and increasing fragility of the vegetation communities on the landscapes targeted for rehabilitation. The story of environmental change in the Eastern Arcs, therefore, must incorporate, in addition to colonial readings, indigenous interpretations of landscape transformation, ecological stability and degradation.

5.3 Related concepts of process

Concepts like *persistence* and *predictability* help to characterize the range and regularity of variation in environmental states over time. *Persistence* describes the natural dynamics of time varying ecosystems while *predictability* describes the expected magnitude of fluctuations. Historians, often reluctant to predict the future, may use these concepts fruitfully to offer planners explanations of past episodes of environmental capriciousness and thus link past changes across landscapes to present and possible future states.

Stability is the propensity of a system to attain or retain conditions of a relatively steady state or stable oscillation. In terms of the historical development of resource management techniques in colonial and post-colonial East Africa, ecosystem policy makers have aimed for stability. However, this very policy trajectory has often led to the diminution of biological diversity and increased the fragility of vegetation communities because it fails to incorporate indigenous interpretations of stable farming systems and local adaptations to them when ecological conditions change. Historical research in East Africa has clearly shown that there are indigenous notions of *stability*. The challenge lies in explaining indigenous history in forums other than those exclusively known to a circumscribed scholarly community.

Resilience is the ability of an ecosystem to maintain its structure and patterns of behaviour in the face of ecological disturbance. It emphasizes non-equilibrium events and processes, variability and adaptive flexibility. Understanding the conditions under which *resilience* breaks down leads to revelations about how Africans have altered past farming and herding systems in response to past stress and crises.

Finally, an analytical approach using these, or other similarly flexible, concepts could inform a policy of *adaptive management*, an approach to resource conservation that embodies a scientific and intellectual commitment to historical ecology. *Adaptive management* should recognize the importance of ecosystem history, the uncertainties in our ability to predict ecosystem behaviour and the desirability of focusing on change and resilience rather than attempting to guarantee equilibrium.

5.4 Evidence for landscape history

Typically, environmental and Africanist history incorporates a number of different types of evidence. Studies of environmental change, in fact, demand such an approach. The Usambara mountains have been the subject of intense scrutiny over the past century and thus lend themselves to a study inclusive of data from oral history, archival records, scientific study, aerial photography, natural history and archaeology.

Oral history

As Leach and Fairhead have shown for social anthropology, complete understandings of ecological change must draw upon oral evidence. Also, Africanist historians have been developing effective methods and interviewing techniques to uncover episodes of ecological stress and crisis in eastern and southern Africa. In the Eastern Arc research, an interviewing strategy should focus on indigenous memories of drought and food shortages. Across the Usambaras, and much of eastern Africa for that matter, severe food shortages resulted in what indigenous peoples remember as *njaa* (Swahili for famine), which records not only an episode of severe hunger, but also an historical benchmark surrounded by evocative stories. Famine episodes are complex, but the narratives associated with them hold clues to demographic change and often point to landscapes under some sort of ecological stress. The stories surrounding particular *njaa* episodes can also illuminate how cultural responses to environmental disturbance and subsequent adaptive responses (such as fine tuning farming systems, marketing systems, trade links, exploitation of new environments) alters a landscape physically and biologically (e.g. Vaughan, 1987; Waller, 1998; Maddox, 1996).

Scientific description and analysis

Another question of evidence arises through an examination of Leach and Fairhead's (1996) rejection of the colonial scientific descriptions of change in West African forest ecosystems. While understanding the conceptual limitations of scientific explanation authored under colonial regimes, they should not be jettisoned *in situ*. By the late 19th century, the particular scientific importance of the Usambaras to the success of colonial agriculture and to the understanding of tropical forests had become clear to the German colonial scientists. After intense lobbying, the German colonial government funded, in 1902, construction on the East Usambara massif of the Amani Botanical Research Station. For the next 40 years, resident German and British scientists produced a body of research that now serves as a foundation upon which to build an interpretation of both scientific discourse and ecological change. In addition to the Amani materials, a bioregional study of the Eastern Arcs could draw upon other scientific literature regarding forests and agriculture.

Archival material

Colonial records from across the spectrum of government agencies and from mission archives have been fruitfully used in environmental histories of Africa. Like the oral evidence, the official colonial and mission sources have great value to analyses of identity and power because they contain veiled discourses that reveal much about the cultural context of colonialism. As records of ecological description, they can, if used carefully, also contribute much to analyses of ecological change because of the value colonial foresters and agronomists placed on natural resource conservation (Conte, 1999).

Archaeological data

As noted in the literature review, archaeological data help link land use practice across space and time. They add the deeper historical framework necessary to understanding culture and land use centuries before written sources became available. At the moment however, the number of sites either already excavated or under excavation in the Eastern Arc mountains is extremely limited. Of particular interest to this study are Iron Age sites, which would indicate settlement patterns for approximately the past 1500 years.

Palynology

Pollen studies allow an understanding of changes in the vegetation patterns essential to the reconstruction of changes in forest composition. Such studies have been used successfully in the Lake Victoria studies cited above.

Images

Several sets of post-World War II aerial photographs are available for Kenya/Tanzania regions containing Eastern Arc massifs. They should be assembled and analysed for changes in vegetation patterns. In addition to the aerial photos, a number of historical maps exist for the Eastern Arc mountains. They can be used in determining old settlement sites and in charting changes in settlement and land use patterns. In addition, photographs of particular landscapes can be used, especially if there was a series of photos taken over time at the same point.

5.5 Tentative research questions

1. What constitutes environmental degradation? What are the differentials of degradation and restoration across the Eastern Arc mountains?

2. What are the dynamic edaphic and hydrological effects of changing land use practices, especially state-run forestry, small-scale farming and mixed farming/herding systems?
3. Under what social, economic, political and environmental conditions do agricultural practices and their landscapes change?
4. What are the essential differences in how different groups with vested interests in natural resources ascribe value to them in particular landscapes? Under what circumstances do these views change?
5. How can ecological history be presented so that it informs policy and development tactics?

6 Conclusions

This chapter argues that historical analyses of ecological change must be firmly rooted in the target environment. Furthermore, if they are constructed with conceptual flexibility, they can inform resource conservation policy. The chapter also strongly advocates a research agenda comparing landscape change in the Eastern Arc mountains of Kenya and Tanzania. The suitability of the Eastern Arcs is based on their similar natural and human histories of forest-based agriculture and herding, but differential rates of ecological degradation both across the mountain range and within the massifs themselves.

The straightforward agenda may obscure the practical difficulties and analytical complexities of evidence collection and its analysis. However, if environmental history is to contribute to the conservation of resilient environments and the restoration of degraded ones, it must present to policymakers and implementers a clear conceptual and analytical picture of environmental change, its causes and effects. Environmental history can in this way present a methodology suitable for interdisciplinary and cross-disciplinary research in a way that informs resource conservation.

References

Collett, D. (1985) The spread of early iron producing communities in eastern and southern Africa: Volume 1. PhD thesis, Cambridge University, Cambridge, UK.

Conte, C. (1999) Colonial science and ecological change: Tanzania's Mlalo Basin, 1888–1946. *Environmental History* 4(2), 220–244.

Crumley, C. (1994) Historical ecology: a multidimensional ecological orientation. In: Crumley, C.L. (ed.) *Historical Ecology: Cultural Knowledge and Changing Landscapes.* School of American Research Press, Santa Fe, New Mexico, pp. 1–13.

Fairhead, J. and Leach, M. (1996) *Misreading the African Landscape: Society and Ecology in a Forest–Savanna Mosaic.* Cambridge University Press, Cambridge.

Feierman, S. (1974) *The Shambaa Kingdom: A History.* University of Wisconsin Press, Madison.

Leach, M. and Mearns, R. (eds) (1996) *The Lie of the Land: Challenging Received Wisdom on the African Environment.* The International African Institute, London.

Lovett, J. (1993) Eastern Arc moist forest flora. In: Wasser, S.K. and Lovett, J. (eds) *The Biogeography and Ecology of the Rain Forests of Eastern Africa.* Cambridge University Press, Cambridge, pp. 33–56.

Maddox, G. (1996) Gender and famine in central Tanzania: 1916–1961. *African Studies Review* 39(1), 83–102.

McCann, J. (1995) *People of the Plow: An Agricultural History of Ethiopia. 1800–1990.* University of Wisconsin Press, Madison, Wisconsin.

Mortimer, M. (1998) *Roots in the African Dust: Sustaining the Drylands.* Cambridge University Press, Cambridge.

Schmidt, P. (1987) Eastern expressions of the 'Mwitu' tradition: early Iron Age industry of the Usambara Mountains, Tanzania. *Nyame Akuma* No. 30, 36–37.

Schmidt, P. (1994) Historical ecology and landscape transformation in eastern equatorial Africa. In: Crumley, C. (ed.) *Historical Ecology: Cultural Knowledge and Changing Landscapes.* School of American Research Press, Santa Fe, New Mexico, pp. 99–126.

Schoenbrun, D. (1993) We are what we eat: ancient agriculture between the Great Lakes. *Journal of African History,* 34, 1–31.

Soper, R. (1967) Iron age sites in northeastern Tanzania, *Azania* II, 19–36.

Sutton, J. (1984) Irrigation and soil conservation in African agricultural history: with a reconsideration of the Inyanga terracing and Engaruka irrigation works. *Journal of African History* 25, 2–41.

Vansina, J. (1995) New linguistic evidence and 'The Bantu Expansion'. *Journal of African History* 36, 173–196.

Vansina, J. (1994–95) A slow revolution: farming in subequatorial Africa. In: *The Growth of Farming Communities, Azania Special Issue,* XXIX–XXX, 15–26.

Vaughan, M. (1987) *The Story of An African Famine: Gender and Famine in Twentieth-Century Malawi.* Cambridge University Press, Cambridge.

Waller, R. (1990) Tsetse fly in western Narok, Kenya. *Journal of African History* 31, 81–101.

Waller, R. (1988) Emutai: crisis and response in Maasailand 1883–1902. In: Anderson, D. and Johnson, D. (eds) *The Ecology of Survival.* Westview Press, Boulder, Colorado, pp. 73–114.

Wasser, S.K. and Lovett, J.C. (1993) Introduction. In: Wasser, S.K. and Lovett, J. (eds) *The Biogeography and Ecology of the Rain Forests of Eastern Africa.* Cambridge University Press, Cambridge, pp. 3–8.

Winterhalder, B. (1994) Concepts in historical ecology: the view from evolutionary ecology, In: Crumley, C.L. (ed.) *Historical Ecology: Cultural Knowledge and Changing Landscapes.* School of American Research Press, Santa Fe, New Mexico, pp. 17–42.

The 'Alnoculture' System in the Ligurian Eastern Apennines: Archive Evidence

16

S. Bertolotto and R. Cevasco

Polo Etnobotanica e Storia, Dipartimento di Storia Moderna e Contemporanea, Università di Genova, Via Balbi 6, 16126 Genova, Italy

An example of the historical analysis of an archival source known as 'Consegne dei Boschi' is presented. This particular historical record allowed the researchers to perform a detailed reconstruction of the local agroforestry system, focused on *Alnus incana* management ('alnoculture') in the 1820s in the upper Aveto valley. The knowledge of ancient practices for modern management problems is underlined.

1 Introduction

As discussed in a previous research paper (Moreno *et al.*, 1998), we have given the name 'alnoculture' to a local agro-silvo-pastoral system – which disappeared in the Apennines at the beginning of the 20th century – that was detected following a circumstantial method ('metodo indiziario') and using a complex of historical sources at a local scale. From fragmentary evidence we have now succeeded in supporting our reconstruction of the cultural cycle involving *Alnus* at the beginning of the 19th century in the upper Aveto valley (Ligurian Apennines, Fig. 16.1). This cycle is compared with a short description dated 1896 produced by the Provincial Forestry Commission that allowed the management of alder woods acting in reply to an application of 'release' of protected land ('svincolo forestale') presented by the local authority of the Santo Stefano d'Aveto village (Croce and Moreno, 1988).

The 'Consegne', seldom expressed in the jargon of the farmers, gives a much more detailed and local picture of the agro-silvo-pastoral practices in

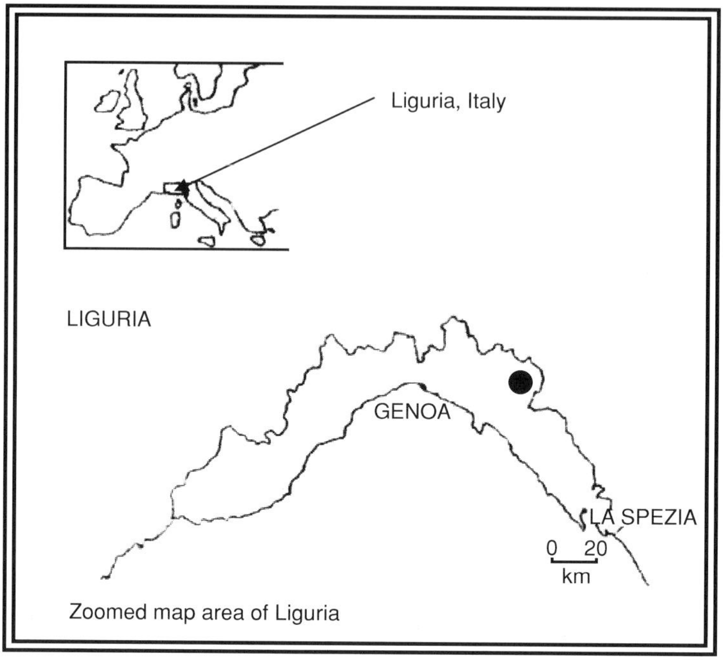

Fig. 16.1. Location of the Aveto valley, in the Ligurian eastern Apennines (NW Italy).

use in the woodlands in this valley during the 1820s. It also brings into question the variability through time of so-called local or traditional practices in forestry.

2 The 'Consegne dei Boschi'

The importance of the 'Consegne dei Boschi' for forest history and historical ecology of the north-western Italian mountain areas has already been stressed in the view of the woodland history of the area (Moreno, 1990). They are an unpublished documentary source produced by the Forestry Administration of the Kingdom of Sardinia in 1822 just at the end of the 'customary regime'. On a local scale we found that every landowner of the nine parishes of the upper Aveto valley, in reply to the first Forestry Administration Act, described his woodland properties at the local judge desk ('consegna'). Every kind of woodland was described except sweet chestnut woodlands because they were assumed – in the text of the Forestry Administration Act – to be equivalent to fields or cultivated land.

In the 'Consegne' the terminology adopted for the different types of woodland and woodland management is local and includes the place name, the location, the neighbouring owners and sometime the size of each parcel of woodland (expressed in 'palmi', 'trabucchi', 'stare' and 'pertiche').

3 Methods and results

This particular type of documentary source describes the tree cover of the area, as it was then, and allows us to know the composition and structure of the woodlands. An electronic archive collected data extracted from 294 'Consegne'; in these documents 1305 different plots of land (real cadastral parcel of wooded land) are described.

Woodland composition, land use and products are arranged in tables in order to detect the local classification system of woodland land use. These completely disappeared with the adoption of the Forestry Law late in the second half of the 19th century (Tables 16.1–16.3).

3.1 Woodland composition

The more frequent tree and shrub species quoted in the 'Consegne' were located in different columns (an example of 20 random items is given in Table 16.1). Archival data were used to find the different tree types in each parcel. The high presence (47%) of alder – pure or mixed with beech or Turkey oak – in these woodlands in 1822 is shown in Fig. 16.2. The rate of presence appears high especially if related to the present decline of *Alnus incana* in the valley.

3.2 Land use

Local land use descriptions, expressed in vernacular terminology, were grouped eliminating the more evident synonyms (for example, 'piantato', 'arborato' and 'popolato' meaning 'planted with . . .'). From a list of 145 different, purely linguistic categories we obtained 60 synthetic categories. An example of this process involving shrubland/woodland terminology adopted in the 'Consegne' is represented in Table 16.2.

By combining woodland composition and land use data we reconstructed the physiognomy and structure of each woodland plot described in 1822 (Fig. 16.3). Comparing the number of plots of land bearing pure alder with plots of land bearing Turkey oak (*Quercus cerris*) and beech (*Fagus sylvatica*), the alder appears largely distributed in a sort of shrubland ('bosco di cespugli, arbusti, costi'), both because of local practices (short rotation coppicing, etc.) and because of the biology of the species. This shrubland stage represents the first and the late stage of the cycle of alnoculture.

Table 16.1. A sample table used for collecting archival data from the 'Consegne' (1822).

Page no.	Doc. no.	Land use	Woodland composition										Note	Products
			cerri	one	fag.	ont.	o.ced.	ont.sel.	one sel.	spini	grat	boc		
666	163	Bosco alberato di arbusti di											Ginepri e costi di faggio	Ricavo legna, pascolo de Bestiami e fogliame
1016	227	Bosco alberato di costi di							1		1			Pascolo per il bestiame
1017	227	Bosco alberato di costi di			1				1					Pascolo per il bestiame
367	84	Bosco alberato di costi di			1									Legna, pascolo e letame
711	179	Bosco alberato di costi di		1									e spine	
350	80	Bosco alberato di costi di												Manutenzione del bestiame, i beni domestici e per ronchi per seminare
570	138	Bosco alberato di piccoli	1											Legna, pascolo e letame
571	138	Bosco alberato di piccoli	1	1										Legna, pascolo e letame
572	138	Bosco alberato di piccoli	1	1										Legna, pascolo e letame

631	156	Bosco alberato di piccoli		1		cespugli e Ginepri	Per legna da brucciare per pascolo e Bestiami o fogliame
632	156	Bosco alberato di piccoli		1		cespugli e Ginepri	Per legna da brucciare per pascolo e Bestiami o fogliame
648	159	Bosco alberato di piccoli		1	1		
1023	229	Bosco alberato di piccoli		1		cespugli, spini e molto grattorato	Per sostentamento loro bestiame, letame, legna
1024	229	Bosco alberato di piccoli		1		cespugli, spini e molto grattorato	Per sostentamento loro bestiame, letame, legna
1117	256	Bosco alberato di piccoli		1			Per legna da bruciare e per pascolo
65	14	Bosco alborato di	1	1		e gratole	Per il bestiame, letame, per chiudere i Domestici, fuoco per la famiglia

Continued over

Table 16.1. *Continued.*

Page no.	Doc. no.	Land use	Woodland composition										Note	Products
			cerri	one	fag.	ont.	o.ced.	ont.sel.	one sel.	spini	grat	boc		
67	14	Bosco alborato di		1						1			e gratole	Per il bestiame, letame, per chiudare i Domestici, fuoco per la famigli
68	14	Bosco alborato di		1	1								Molto gratorato	Per il bestiame, letame, per chiudere i Domestici, fuoco per la famiglia
383	89	Bosco alborato di	1										e qualche arbusto di one	E gratorato servente per il fuoco
384	89	Bosco alborato di											qualche cespuglio di one	Servente per il fuoco

Cerri (*Quercus cerris*) – Turkey oaks; one, 'one' (*Alnus incana?*) – alder; fag., 'faggio' (*Fagus sylvatica*) – beech; ont., 'ontano' (*Alnus incana?*) – alder; o.ced, 'ontano ceduo' (*Alnus incana?*) – alder; ont.sel., 'ontano selvatico' (*Alnus* sp.) – alder; one sel., 'one selvatiche' (*Alnus* sp.) – alder; spini, 'spini' (*Crataegus monogyna, Prunus spinosa?*); grat, 'gratole' (*Rubus ulmifolius, Rubus hirtus, Rubus idaeus?*); boc, 'bocchi' (*Crataegus monogyna, Prunus spinosa?*).

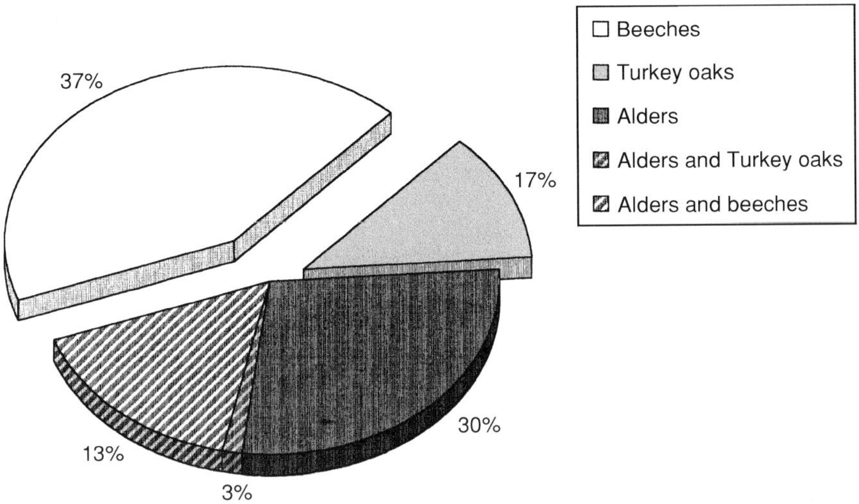

Fig. 16.2. Presence of alders (*Alnus incana* (L.) Moench) in the woodlands of the Aveto valley (data elaborated from the 'Consegne', 1822).

Table 16.2. An example of combining categories of land use from local land use terminology.

Local land use terminology	'Synthetic' categories of land use
Bosco alberato di costi di	Bosco
Bosco alberato da alberi novelli di	Terra arborata
Bosco alberato di arbusti di	Bosco di cespugli, arbusti, costi
Bosco alberato di cespugli di	Terra prativa, pascolativa e coltiva
Bosco alberato di piccoli	Alberi
Bosco alberato di scateri ossia arbusti di	Pascolo arborato
Bosco arborato di grappuli di	Terra con costi, cespugli, arbusti
Bosco con cespugli e arbusti di	
Bosco di costi di	
Bosco ed arbusti	
Bosco grattorato alberato di cespugli di	
Piccoli boschi	

When alder is mixed with Turkey oak we found 'terra prativa' as the commonest land use term employed in describing the parcel; the same relationship appears for each mention of pure Turkey oak woodlands. The alder, through increasing soil fertility, helped *Quercus cerris* in leaf fodder ('frasca') and/or wood production. This co-plantation is occasionally expressed in the documents as 'cerri onati'.

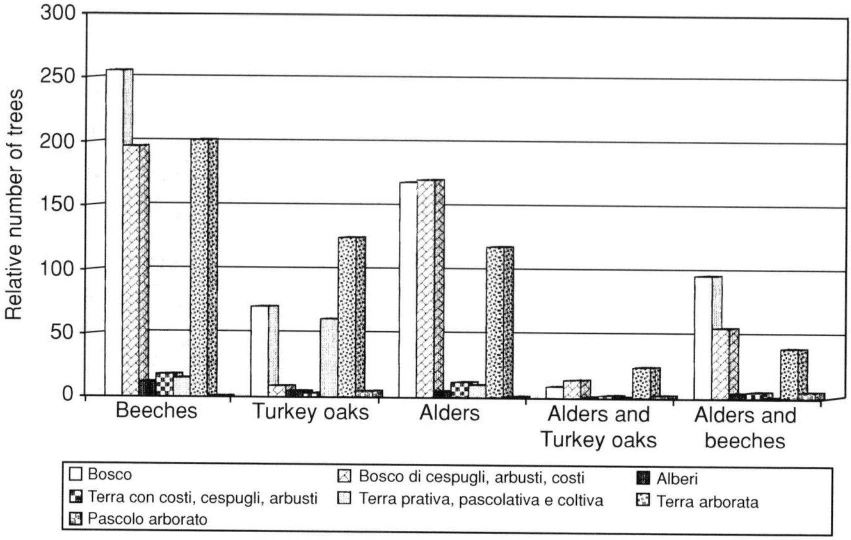

Fig. 16.3. Frequency of land use and woodland composition in the Aveto valley (data elaborated from the 'Consegne', 1822).

3.3 Products

For the 967 parcels of land bearing alder (pure or mixed), the main types of produce were closely examined (Table 16.3). In these alder woods a multiple use of resources was at work, involving mainly slash-and-burn agricultural crops ('Ronchi'), pasture, firewood and vegetation manure obtained from collecting leaves ('. . .e in quello popolato di ontano si fanno dei Ronchi e come pure si raccoglie foglia per il concime'; Archivio di Stato di Genova, doc.197).

The 'ronco' practice is frequently quoted in documents describing alder woods and is often connected to 'chiudere le terre domestiche' (to enclose cultivated land) that could mean a temporary protection of 'seminativo' (rye and oats cultivation) from grazing (Moreno, 1984). The enclosure is made using the alder itself (Piussi, 1988) or thorny shrubs (*Crataegus monogyna, Prunus spinosa, Rubus* spp., *Rosa canina*) which sprout after a confined fire ('abbruciamento del soprassuolo') as used in the 'ronco' practice.

Comparing woodland composition, land use and products we reconstructed, from fragmentary data, the cycle of alnoculture and its considerable variations.

4 Reconstructing the 1822 alnoculture cycle

The alnoculture cycle reconstructed previously using a document of the Forestry Administration dated 1896 (Moreno *et al.*, 1998) is now compared with our data (Fig. 16.4).

Table 16.3. An example comparison of products, woodland composition and land use used to construct the 1822 alnoculture cycle.

Land use	Products	Woodland composition
Bosco alberato di arbusti di		
1	*Gratorato, pascolo e fuoco*	One e faggio
8		One
1		One selvatiche
7	Pascolo e per chiudere le terre selvatiche	One e faggio
0	(Vuote)	Ontani cedui
Bosco alberato di cespugli di		
3	A nulla	One
1		One selvatiche
2	*Gratorato, pascolo e fuoco*	One
2	Legna, foglia e pascolo	One, one e faggio
24		One selvatiche e ontani
1	Pascolo e chiudere le terre domestiche e fuoco	One selvatiche
1	Pascolo e fuoco	Ontani selvatici
3	Pascolo e per chiudere le terre selvatiche	Ontani
Bosco alberato di costi di		
2	*Bestiame, foraggio pascolo fuoco segale*	One e cerro, one e faggio
2	Pascolo	One
3		One selvatiche
1		Ontani selvatici
9	Pascolo e fuoco	One

> Si sogliono nei territori medesimi fare dei ronchi segnatamente ove sono le One facendo colla legna che si taglia delle fornaci per ingrassare il terreno, e nel sito tagliato e roncato si semina per due tre e talvolta quattro anni. Indi nel corso di qualche anni successivi il sito roncato e seminato ritorna nel primierio suo essere massimamente ove sono le One. (doc. 44)

Fragmentary descriptions – as the example quoted above shows – reflect a variability in the span time of the different practices involved in the cycle (Fig. 16.5). Many factors may produce this variability, such as land morphology, soil resources availability, livestock caring capacity, individual strategies, etc. The cycle was employed in the alder woods, as well as in the mixed woods with alder, beech and Turkey oak.

The felling of alder – stage one – might occur at from 3 to 12 years and the cycle period was related to when the wood could be utilized again for fire and fodder. Alderwood is specific for the practice of 'ronco': the wood was used to make 'fornaci' (confined fire) – stages two and three – and the ashes were

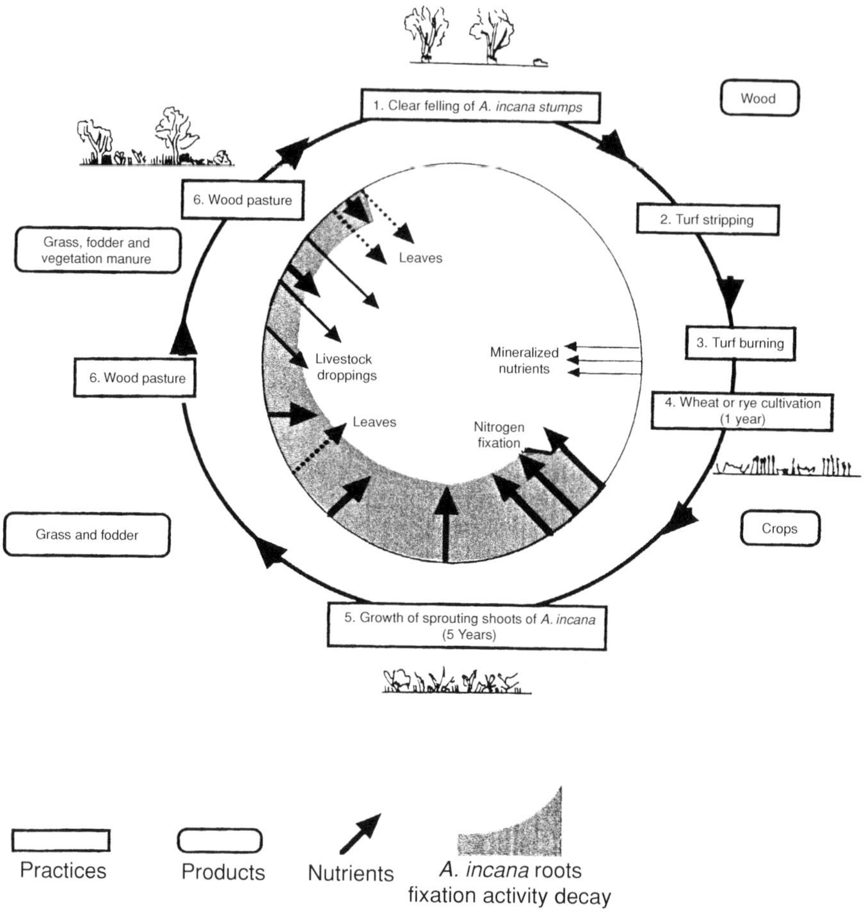

Fig. 16.4. The 'alnoculture' cycle according to the 1896 prescriptions (Moleno *et al.*, 1998).

spread over the soil to manure it. It is probable that 'ronco coperto' helped to save useful fruit trees such as cherry trees ('ceraso') and other leaf-fodder trees such as Turkey oaks ('cerro') and ash ('frassino'); big trees of these species survive in the historical sites of 'alnoculture'.

The rye cultivation (generally called 'terra seminativa' or 'seminerio') in the 1822 documents usually lasted much longer (from 2 to 5 years) than in the cycles reconstructed for the same valley in 1720 and 1896 (Moreno *et al.*, 1998):

> ... l'uso di questo Bosco del quale l'alberatura serve a nessun uso, meno per abbruciiare, serve al pascolo. Ogni 3 anni si può seminare, ed il seminerio dura tre anni, e si riprende quando il bosco è cresciuto al punto per fornire la legna per l'ingrasso del terreno abbruciandola. (doc. 150)

Wood grazing of sheep, goats and, rarely, cattle is allowed after the last agricultural crop, when it was not an obstacle to the growth of the young trees.

Confined fire ('fornaci') in the alnoculture cycle helped to clear the undergrowth from the young and plaited hawthorns ('bocchi') and particularly from brambles ('gratole') forming the temporary hedges. The oldest 'bocchi'

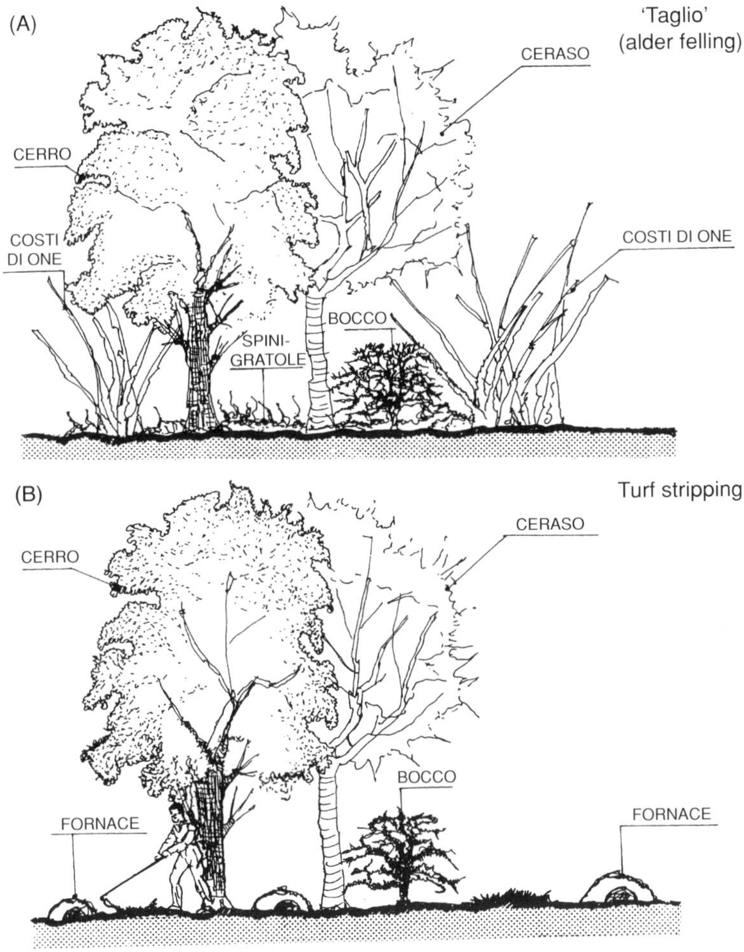

Fig. 16.5. (*and over*) The prevailing practices and local terminology in the alnoculture cycle according to the 1822 'Consegne'. The cycle could last from 3 to 12 years. The stages of this cycle were reconstructed from the composition, land use and products documented in 1822 (an example is shown in the grey area in Table 16.3). The alnoculture cycle (in 1820): (A) alder felling; (B) turf stripping; (C) turf burning; (D) rye and oats cultivation; and (E) wood pasture.

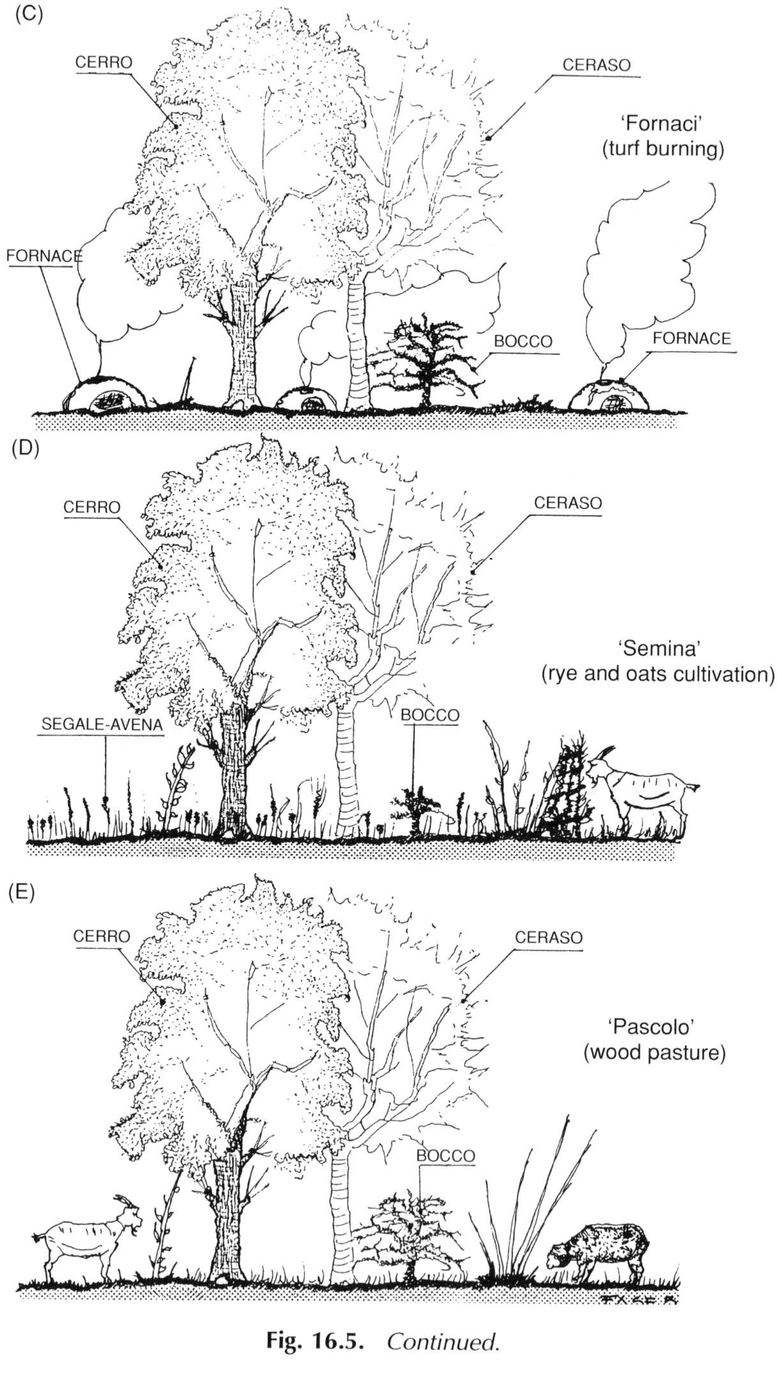

Fig. 16.5. *Continued.*

were preserved by this practice and were used to protect the 'seminerio' from the surrounding grazing area.

5 Conclusion

Environmental history research focused on forestry resources in the Mediterranean mountains should be directed to a basic group of questions concerning the historical reconstruction of local practices and land use systems. The alnoculture system in the Apennines is an example of these practices which has disappeared during the last 150 years of modern forestry management and plantation.

This approach allows us to develop other historical topics, and, using this new approach and unexploited sources (in archives and fields), to examine further the history of rural society and local nature knowledge. Environmental archaeology experiments involving historical practice reconstruction and their effects on present alder population could be planned in order to:

- check the effects of these practices which have disappeared, and recognize indicator species in local flora;
- understand the historical behaviour of alders in the local ecological system;
- assess in a quantitative way the historical economic value of the system.

The reintroduction of the whole alnoculture system (or of some stages such as grazing and coppicing) could be useful in managing and planning alder woods presently degraded. What kind of management can we foresee for the historical alnoculture sites? In the upper Aveto valley – today a regional natural park – in particular, alder woods should be conserved for their historical environmental value and as a potential game resource. A number of actions are presently planned ('Piani di miglioramento ambientale per l'incremento della fauna stanziale e migratoria' – Provincia di Genova) in order to increase the traditional shooting game (such as hare and partridge) in the valley. A special plan is devoted to recovering ancient cultivated areas by sowing grain crops such as *Secale cereale*, *Avena sativa*, *Fagopyrum aesculentum* and *Hordeum* sp.

The adoption of the alnoculture systems in this plan would be of interest in three main areas:

1. Conserving the alder population.
2. Producing grain crops for game.
3. Increasing rearing resources, especially for the breeding of the local historical bovine animal called 'Cabannina', which is particularly adapted to the Apennines woodland/shrubland habitats.

References

Documentary Sources

Archivio di Stato di Genova (xxxx) *Prefettura Sarda, Boschi e Foreste, S. Stefano d'Aveto – Consegne dei Boschi.* pacco 207. Archivio di Stato di Genova.

Archivio di Stato di Genova (xxxx) *Prefettura Italiana – Verbali della seduta del Comitato Forestale di Genova.* Vol. V 1894/1905, Seg. 28 VII 1896. Archivio di Stato di Genova.

Bibliography

Bertolotto, S. (1997) *Storia e copertura vegetale nell'Appennino: effetti delle coltivazioni temporanee in età storica in val d'Aveto.* Tesi di laurea in Geografia storica dell'Europa. Università di Genova, Genova.

Croce, G.F. and Moreno, D. (1988) The geographical effects of the forestry law 20/VII/1877 in Liguria (NW Italy). In: Salbitano, F. (ed.) *Human Influence on Forest Ecosystems Development in Europe.* Pitagora, Bologna.

Daniere, C., Capellano, A. and Moiroud, A. (1986) Dynamique de l'azote dans un peuplement naturel d'*Alnus incana* (L.) Moench. *Acta Oecologica* 7, 165–175.

De Notaris, J. (1844) *Repertorium Florae Ligusticae.* Torino.

Gismondi, A. (1950) *Prospetto della flora ligustica. Guida botanica della Liguria.* Scia, Genova.

Guidi, M. and Piussi, P. (1991) The influence of old rural land-management practices on the natural regeneration of woodland on abandoned farmland in the Prealps of Friuli, Italy. In: Watkins, C. (ed.) *Ecological Effects of Afforestation. Studies in the History and Ecology of Afforestation in Western Europe.* CAB International, Wallingford.

Moreno, D. (1984) The agricultural uses of tree-land in the north-western Appennines since the middle ages. In: *History of Forest Utilization and Forestry in Mountain Regions.* Symposium der ETH Zurich.

Moreno, D. (1990) *Dal documento al terreno. Storia e archeologia dei sistemi agro-silvo-pastorali.* Il Mulino, Bologna.

Moreno, D., Cevasco, R., Bertolotto, S. and Poggi, G. (1998) Historical ecology and post medieval management practices in alderwoods (*Alnus incana* (L.) Moench) in the northern Apennines, Italy. In: Kirby, K. and Watkins, C. (eds) *The Ecological History of European Forests.* CAB International, Wallingford, pp. 185–201.

Penzig, O. (1897a) Flora popolare Ligure. Primo contributo allo studio dei nomi volgari delle piante in Liguria, Atti Soc. Ligustica Sci. Nat. Geogr. *Genova* 8, 249–308 and 341–379.

Penzig, O. (1897b) Florae Ligusticae Synopsis, *Ann. Mus. Civ. Storia Nat. Genova* 38, 423–531.

Pignatti, S. (1982) *Flora d'Italia,* Vols I, II, III. Edagricole, Bologna.

Piussi, P. (1998) Piantagioni di ontano nero in prati falciabili nel Friuli orientale. *SM Annali di San Michele,* no. 11.

Between Nature and Culture: the Contribution of Anthropology to Environmental Study

17

P. Clemente[1], N. Breda[2] and V. Lapiccirella-Zingari[3]

[1]*Docente di Antropologia Culturale, Università 'La Sapienza' Roma, Dipartimento di Glottoantropologia, c/o Dipartimento Filosofia, Via Carlo Fea 2, 00161 Rome, Italy;* [2]*Dottorato in Scienze Etnoantropologiche, Università 'La Sapienza' Roma, Dipartimento di Glottoantropologia, Vicolo Sère 17/a, 31020 San Vendemiano, Treviso, Italy;* [3]*1 rue Argentina, 73100 Chambery, France*

Anthropology has provided an important contribution to knowledge about relationships between man and environment, both on the level of the variety of practical experiences and cultural uses of the natural world, and on the level of the strong symbolic, linguistic and metaphoric bond between man and nature. More than ever the anthropological method, which elaborates its accounts and documents in close connection with the populations and environments investigated, can today show the variety and complexity of those relationships, which are updated in contemporary contexts through memory, oral tales and words. Local forms of life, as accounted for by the anthropologist, can be profitably compared with other disciplines. Furthermore, anthropology can help improve management of today's environment and can increase the likelihood of a better and more balanced future. Two cases are considered, one relating to a wood and one to bogs and grasslands (Tuscany and Veneto, Italy, respectively), which use anthropological methods and disciplinary concepts, focusing on qualitative data.

The king (to Macbeth): Welcome hither: /I have begun to plant thee, and will labour/ To make thee full of growing.

(to Banquo): Noble Banquo, /That hast no less deserv'd, nor must be known/ No less to have done so, let me unfold thee, /And hold thee to my heart. Banquo There if I grow, /The harvest is your own.

(W. Shakespeare, *Macbeth*, Biblioteca Universale Rizzoli 1980, p. 60)

1 A community of life between populations and their environments (P. Clemente)

Italian anthropological studies, that have been recognized under the definition of demo-ethno-anthropological cultural heritage,[1] have contributed important knowledge on the forms of relationships between man and environment, both in terms of the variety of experiences and cultural uses of the natural world, and of the strong symbolic, linguistic and metaphoric bond connecting people with the natural world.

Long before the 1960s, when the fields of 'human ecology' and 'cultural ecology' were defined,[2] anthropology using travellers' and missionaries' descriptions had concerned itself with the great diversity of relationships between populations and environments. It had developed some of the basic concepts of the discipline, such as 'taboo' and 'totem'. James Frazer's great work, *The Golden Bough*, published at the end of the 19th century, is to a great extent a comparative and unified interpretative inventory of the relationships between nature and primitive worlds. The book, which is set in a wood and depicts a rite of the classical world, is largely dedicated to the cult of trees, to the rebirth of vegetation and assumes wood–vegetation–trees as fundamental pivots of human representations.[3] In the same period the study of folk traditions (later called demology) mainly consisted of investigating the repertoires of folk zoology, botany, and so on. This branch of studies has continued gathering information on the uses of herbs, techniques and rituals that define local modalities of relating to nature.[4]

When scientific or modern anthropology developed in the 1920s and 1930s, the themes of the relationship between cultures and the natural world remained central, both in ethnography and in the emerging interpretative theories. It was especially the position of Lévi-Strauss in French culture and of ethnomethodology in the American area (1960s and 1970s) that gave crucial theoretical innovations to what appeared to be a descriptive catalogue or a positivistic interpretation of archaic and surviving elements. These new horizons, partly interlinked in an approach of 'cognitive anthropology'[5], aim at understanding the local logic underlying classifications, experiences and names, that recognize a 'native knowledge' of the natural world. This form of knowledge corresponds to the 'savage thought' theorized by Lévi-Strauss: an original, practical and universal thought.[6]

The studies of history of material culture, dating back to the 1970s, focused anthropological attention on techniques, rural landscapes, methods of cultivation and nature knowledge, developing a deeper bond between anthropology and history.[7] Consequently, forests, woods and vegetation environments are today a favourite subject of anthropological research; its attention is always focused on knowledge and uses, values and symbols, that make nature a fundamental part of the world of man.[8]

Ecologists' and bioregionalists' efforts have been affected by some anthropological themes and research. Anthropology has shown that many civilizations assigned sacred features to the man–nature relationship. Furthermore, on the horizon of contemporary ideologies, anthropology is a source of survival techniques and techniques of sustainable life, so that local culture experiences become repertoires for a more balanced future.

Anthropology applies the concept of 'fundamental company' or 'horizon of life' which was emphasized by Ludwig Wittgenstein in a controversy (though not an irresolvable one) with Frazer's position. He wrote:

> It can't have been a worthless reason, in fact there can't have been a reason if some populations used oak-trees (in their rituals); it was simply the fact that those populations and the trees were united in a community of life, and were therefore close to each other not by choice, but having grown up together ... We could say it was not their union to give the opportunity for these rituals, but rather their separation.[9]

Anthropology is proud of its contribution to understanding the diverse human capacity to adapt, to act with and to conceive the environment. It interprets the experiences of different peoples no longer as 'archeology' or 'backwardness', but as an inventory of still thinkable and possible alternatives. This is in contrast to scientific monism brought about by modernism and the idea of a winning, irreversible, unilinear progress. It is indeed this ideology of winning science that appears backward and superstitious; not only from an anthropological point of view.[10] Anthropology is also proud of applying the modality of the man–nature relationship as a central theme, while it feels unrelated to a renewed causalist and determinist ecologism. Anthropocentrism is not (and for most anthropological experiences it has not been so) only destructive but can be brought back to a polyphonic dimension.

Contemporary anthropology has resumed an old dialogue with the arts, especially literature. Thanks to that dialogue it has realized how deeply aesthetic and symbolic experiences are marked by the community of life with the world of trees. It is the primary world of man – in language, in metaphors, dreams and memory. It is to this dialogue with literature, in particular with poetry, where the many-sided human feeling of being-in-the-world is directly caught, that we have dedicated some 'epigraphs' taken from Italian poems. We do not regard them as ornaments, but as direct ways of grasping the human feeling of natural landscapes, where life and thoughts on life have their foundations.

2 Notes for a historical anthropology of local forest environments (P. Clemente)

> ... and the wing, the skies, the woods, the song: /old woods with a leaf falling,/ mute, whenever the heart beats. (G. Pascoli, Tristezze)[11]

Extra-European ethnology, the history of folk traditions (demology) and cultural anthropology have paid particular attention to the local dimension, 'to the village', as well as to the comparative dimension for specific cultural 'traits'. The local dimension has proved to suit our studies perfectly, even when combined with the comparative dimension for cultural traits, because it avoids considering single villages as 'typical models' of environments or populations. Clifford Geertz, the main representative of the interpretative trends in American anthropology, has stressed that anthropologists study 'in villages' and not 'villages',[12] meaning to say that they meet people, listen to stories, reconstruct images and observe contexts, but do not claim to see the essence of village life (a sort of radiography of what is invisible to the inhabitants). Nor do they deduce aspects of the group or the people. Anthropology is therefore proposed as an interaction, leaving traces to be interpreted in a sensible way, so that elsewhere somebody else can understand and imagine what the anthropologist has understood.

The interweaving with historical work in the local dimension and the central role assigned to the performances of 'social actors' helps anthropology avoid any conceited scientism and to reconstruct voices, traces of knowledge and pictures of local forest environments as seen by people who experienced them as a part of their own community.

This chapter briefly presents two cases related to the above-mentioned interests, located in the Italian woodlands. The case studies are not described in an analytical or systematic way. The authors offer synthetic considerations on the nature of these local worlds and the method of approaches used in both cases.

The first case is set in Tuscany, in a deeply changed environment (Zingari), while the latter case has been investigated in the Veneto region (Breda), in the framework of a strong political negotiation for environmental protection. In both cases, the strategic resources for anthropological knowledge are based on 'oral sources', i.e. the natives' memories, stories and representations.

2.1 The memory's wood: 'In winter there were people in the woods' (V. Lapiccirella-Zingari)

> ... Far away from our eyes live the woods /enclosed with ancient words, ruins of past times,/ they live in places where we no longer are.
>
> (F. Fortini, Una facile allegoria. The text of the poem is at the entrance of the Museum of the Wood, Orgia)[13]

The wooded area of Orgia, a hamlet in the municipality of Sovicille, in the Province of Siena, is a wood of mixed oak groves with evergreen bushes, chestnut groves and riparian vegetation on the Merse river; it also has conifer plantings due to the degradation of the mixed woods. There is a complex ecosystem consisting of wood, underwood and river. This is a picture of what the vegetation at Orgia was like, of what remains and what no longer exists. In terms of community use, the wood does not exist any more. It lives far away from our eyes, in a place where we no longer live, in the memory of those who inherited it, used it, took care of it and lived on it. The wood still lives in the memory of a complex and mobile group of workers (charcoal burners, woodcutters, woodgatherers, woodmen, tenants) and of the sharecroppers' families who lived permanently in farms until the 1960s and characterized the commonest agrarian management in the Sienese area. That was, in fact, the sharecroppers' wood.

Starting from that memory, oral tales update a complicated world of competence, knowledge, mastery of spaces, rhythms and times, changing according to the social figures taken into consideration. The stories do not concern only Orgia's museum of the wood, which spurred us on to the research,[14] they concern all of us; being the reconstruction of a 'type of wood', of a 'system of knowledge and experience', an environment of man, they help us to understand the various ways of life in various environments and perhaps to build projects for the future.

The wood was one of the manifold resources in a 'multiactivity' involving sharecroppers.[15] However integrated, these resources were strategic. They supplied acorns, mushrooms, big and small game, firewood and chestnuts; the wood was a strategic non-agricultural resource for Tuscan sharecroppers living in environments poor in plains and fertile land. The wood was also an essential resource for woodmen, charcoal burners and woodcutters who did not live in one place; they moved locations looking for work and were employed seasonally. The wood marked the rhythm of life in a community living nearby or even inside it (there were farms in the wood, built within its area). The rhythm marked meetings and separations, but also determined the closeness of different destinies and shared language. The feast sealed this pact of life spent around a common vegetative environment. People would meet, eat together on various occasions and barter products. Sharecroppers would sometimes work as woodsmen in winter, and woodsmen would sometimes work in summer in agricultural activities. To all groups of rural society – peasants and woodsmen, sharecroppers and workers, small and big landowners – the wood appeared to be a reserve of products and an opportunity of employment essential for local economy.

Although it was based on a close familiarity, the relationship with the wood maintained some features of fertile ambiguity, as it was a land where a symbolic language arose and developed, which, in connection with old times, was reconstructed and transmitted by the community. The wood was a space of freedom and individual expression, elsewhere impossible. It was the place of imagination 'par excellence', where improbable things could be possible with extraordinary meetings, wonderful harvests, memorable hunts and frequent exceptional visions; the wood, which was not completely familiar, hinted at new dimensions beyond daily life.

In the peasants' and artisans' memory the wood was intensely experienced and known. It was also intensely used, but in a way that conserved resources. In their perception, the ancestors managed the wood with an implicit respect for culture and nature, a relationship full of beliefs and stories.[16] Today's voices have inherited the ancestors' voices and recall therefore the voices of the past life of the woods. If we listen to their testimony we are invited to follow the ancient and future paths of what is now called 'sustainable management'.[17]

What an ordinary visitor may nowadays consider merely a beautiful wood, 'smiling' according to the synaesthetic expression of a guidebook, the ancient dwellers, who grew up together with it, saw it as a monster, partly wild and partly tame and 'escaped' from any control. The richness of toponyms, almost inch by inch, opposes the non-knowledge of the Sunday townsman who 'meets' the woods for his walks and for mushrooming; he shares a new community with the natural world.[18] But this (today's) relationship is less close and conscious, since it has lost the ancestors' heredity. Non-knowledge brings about the need of a museum with botany itineraries in the open air: 'this is called heather, this is called holm oak'.

Meanwhile, old wood dwellers feel that the wood is going to ruin.[19] This is also the end of a community. The feeling of not belonging grows and deepens, investing whole places and landscapes. Bushes obstruct paths and cancel the spaces opened by the countless presences and transits of the past. Today the wood is described as a lost order. The present is the opposite of that order.

Memory, however, opposes oblivion and, in addition, the oldest uses of the resources, such as hunting and gathering, are coming back, though in new forms. The composition of the population has changed and citizens are coming back to the country after abandoning it. In this context, strong connections to the wood and its resources are being confirmed again. Nowadays in autumn a varied population, local and urban alike, goes to the woods. Mushrooming is maybe the most significant sign of a relationship which refuses to come to an end. It is to this new world of close or distant tourists, ignoring old toponyms, that we suggest listening to the voices of the past. If townspeople care for the wood, they may later discover the distant voices and the ancestors' world. Knowledge needs love to commit itself to the choices of a new life. Anthropology does not urge these choices, yet it does hope that its knowledge may favour them.

2.2 Palù: the diversity of woods and wetlands (N. Breda)

> The branch, the real branch reaches / the most covert enclosures of water, leaning forward, / always greener than its not-reaching / Protect the sweet shrewdness of the grass / from the careful order of the crops / of their bright green sword-like spikes / where summer is so sharply visible / You tender and present everywhere / crowds of surprises / you talk so intimately / increasing your understanding . . .
>
> (Personal communication from Andrea Zanzotto to N. Breda. Now in *Il Gallo Silvestre*, n.12, 2000, Editrice 'I Mori', Siena, Italy, pp. 47–48.)[20]

The palù (bogs) of the Quartier del Piave area in the Province of Treviso,[21] are clayey grounds, laid on the line of resurgences, characterized by floods, swamps and springs of water with meadows, bushes and here and there a wood (of middle and high trunk) working its way into the cultivations. It is a lush world, but always on the margin of agriculture and regarded as wild. The palù are a very vital habitat, too: an authentic biological reserve to be protected. This territory is at the same time opposed and complementary to the fields. It participates in the landscape construction by man through gradual processes of adaptation and modification. It remains distant and useful at the same time. Since the palù, unlike the sharecroppers' wood, are felt as alien and not as familiar, they represent a different typology both of woodland and of 'growing together' and of community between people and nature.

The soil is thick, cold, clayey, virginal (in the local idiom), virginal as not fecundated and non-arable.[22] Rather than a family member, the palù appear to be a stranger (maybe G. Simmel's 'internal stranger' in a vegetative version) in whose eyes it is possible to understand one's own diversity.[23] This diversity is locally represented in terms of a feminine image. The feminine element is not familiar. It is alien, unfertile, wild. This connection gives rise to classificatory principles for naming plants related by analogy to sterile animals like oxen and mules. Around the palù, 'good to think', a reflection in the rural world develops, which is interested in the categories of sexuality, fertility as well as reproduction and which connects man, earth, plants and animals in a particular 'local eco-cosmogony'.[24] This local way of thinking is therefore based on images and metaphors of vitality and life, seen as the primary elementary process of reproduction. Owing to the peculiarity of this non-arable land, whose wildness also contains a primary idea of Eden, its products result from gathering rather than cultivating. People who appropriate them show the mentality of a primitive hunter–gatherer rather than of a peasant. The palù do not require labour, do not depend on rain, do not need any supplications, or blessings, crosses or prayers. They therefore possess a nearly self-fertilizing principle, as if they were an Eden or a land of plenty. They have a primigenial dimension, preceding any original sin and are consequently both alien and complementary to the peasant community that appropriates them.[25]

The metaphor of the palù as a primordial Eden, however, must reckon with the dominion of damp and water, a difficult and always unstable environment.[26] Non-dominated water is the enemy of 'rural progress', of complete reclamations, of the modern trend to make everything uniform. These lands are threatened by projects of water drainage that aim at obtaining cultivatable soil. A second threat consists of motorway projects which profit by the 'marginality' of the palù.[27] Peasants are still in touch with this diversity of spaces and their memory 'thinks of' and 'tells about' the slightly disquieting diversity of the palù in a discontinuous connection with the beloved ploughed land. At the same time they are aware of living in a world offering an additional mysterious chance.

It is in this context that the palù, threatened by 'taming' and 'yoking' to ploughs and motorways, resist, passively but with strength. They resist because they represent a challenge. The palù's earthly and aquatic otherness relates to hunter–gatherers' inner experiences, to a pre-agricultural world, to a self-sufficient land recalling a primordial myth about feminine parthenogenesis of life. Becoming a symbol of biodiversity, the palù gather intellectual and even poetical forces in support of their protection.[28] For a number of reasons, or maybe for no reason (as Wittgenstein could have said) this environment of man is often felt as a crucial issue for a future which should be able to respect the unquiet foundations of life safeguarding their conceivability instead of destroying them.

We have given two examples of Italian research based on the same methods of work and which provide accounts of historical forest environments seen in the perspective of local life. Our purpose was to demonstrate the potentialities offered by a method of research and by disciplinary concepts which confront historically more-consolidated fields of knowledge. Gregory Bateson made an attempt to be a bridge between different disciplines and different forms of knowledge. He praised artists and visionaries for their deep understanding of the world. We feel that the disorder of the palù would have been in perfect agreement with his concept of 'mind and nature'. In addition the sharecroppers' wood, with its variety of activities and rhythms, could have been an emblem of how a communication system between nature and culture works. Bateson, in fact, thought that the richest knowledge of the tree includes both myth and botany. His daughter, on editing his posthumous work *Dove gli angeli esitano*, wrote that the final words of the book could have been William Blake's verse 'May God us keep / From Single vision and Newton's sleep!'[29] To the poet's prayer we add these short presentations of worlds made up of memory and words, emerging from local experiences. They are the concrete objects of our anthropological passion for the diversity in life.

Notes (N. Breda)

1 For an updated reflection on the demo-ethno-anthropological cultural heritage, starting from the experiences of demological studies, see Clemente, P. (1996) *Graffiti di museografia antropologica italiana*. Protagon Editori Toscani, Siena. For a comprehensive, detailed discussion on the studies of demology, folklore, folk traditions (and a precise definition of the meaning of these terms) see Cirese, A.M. (1971) *Cultura egemonica e culture subalterne. Rassegna degli studi sul mondo popolare tradizionale*. Palumbo editore, Palermo.

2 For a synthetic introduction to human ecology and cultural ecology see Barrau, J. (1981) Ecologia. In: Cresswell, R. (1981) *Elementi di etnologia*. Il Mulino, Bologna, pp. 77–90. The suggested bibliography mentions major American scholars, such

as R. Netting, M. Salhins and R.R. Rappaport, and French authors, such as G.E. Bourgoignis and J. Barrau. More recently, the first chapter in Viazzo, P.P. (1990) *Comunità alpine. Ambiente, popolazione e struttura sociale nelle Alpi dal XIV sec. ad oggi* (Il Mulino, Bologna) discusses scientific results and the contributions given by these studies. Remarkable bibliographical indications are contained in Solinas, P.G. (1991) *Itinerari di letture per l'antropologia. Guida bibliografica ragionata*, Ed. CISU, Rome. Also see the important critical bibliographical excursus in Orlove, B.S. (1980) Ecological anthropology. *Annual Review of Anthropology* 9, 235–273.

3 The myth of silvan Diana and of her cult in Nemi holy woods is dealt with by J. Frazer (1965) in the first chapter of his classic *Il ramo d'oro della magia e della religione (The Golden Bough)*, Vol. 1, Editore Boringhieri, Torino, pp. 7–18. The volume also investigates the cult of trees and its influences in modern Europe (pp. 193–215).

4 As for this type of work, in Italy, there were studies by G. Pedrotti, V. Bertoldi, O. Penzig in the early 20th century. In France, the great ethnographic survey 'Les végétaux dans le folklore et l'ethnographie' was organized by A.L. Mercier and published in *Ethnographie* in the years ranging from 1940 to 1960. E. Rolland compiled an inventory in 11 volumes, *La flore populaire ou Histoire naturelle des plantes dans leurs rapports avec la linguistique et le folklore*, published between 1896 and 1914. A discussion on the history of the studies is contained in G.R. Cardona (1985) *La foresta degli studi. Manuale di etnoscienza*. Laterza Bari. A survey of the passage from these studies to a more modern anthropology of knowledge is in Breda, N. (1997) Figure sociali dei saperi naturali. *La Ricerca Folklorica* 36, 111–131, Grafo Editioni, Brescia, Italy.

5 A fundamental reference for the classic period of ethnoscience is by Tyler, S.A. (1966) *Cognitive Anthropology*. Academic Press, New York. For an interesting development of cognitive anthropology see Boyer, P. (ed.) (1993) *Cognitive Aspects of Religious Symbolism* (Cambridge University Press, Cambridge), in particular the essays by S. Atran, P. Boyer and C. Severi. A recent overview is in D'Andrade, R. (1995) *The Development of Cognitive Anthropology*. Cambridge University Press, Cambridge.

6 We are referring in particular to that 'science of the concrete', whose nature of 'bricolage intellectuel' is investigated by Lévi-Strauss, C. (1962) in *La pensée sauvage*. Librairie Plon, Paris, pp. 12–49.

7 As for aspects more directly related to forest history we refer to the debate opened by the Italian journal *Quaderni Storici* in two monographic issues, *Boschi, storia e archeologia* 49 (April, 1982) and *Boschi, storia e archeologia* 62 (August, 1986). For a contextualization of the woods as a component of rural landscapes and a methodological discussion on the necessary integration between fieldwork and archives, testimonies of material culture and complexity of historical processes, which determine the dynamics of agricultural, forest and pastoral systems, see Moreno, D. (1990) *Dal documento al terreno, Storia e Archeologia dei sistemi agro-silvo-pastorali*. Il Mulino, Bologna. Another important work is Ambrosoli, M. (1992) *Scienziati, contadini e proprietari. Botanica ed agricoltura nell'Europa Occidentale, 1350–1850*. Einaudi, Torino. Still fundamental is the essay by Cole, J.W. and Wolf, E.R. (1974) *The Hidden Frontier: Ecology and Ethnicity in an Alpine Valley*. Academic Press, New York. As regards the rich extra-European literature we would like to mention, at least, an essay characterized by a particular interweaving of history and ethnology, of research in archives and fieldwork; we are referring to N. Wachtel's 1990 masterpiece *Le retour des ancêtres. Histoire régressive des Indiens Urus du Bolivie*. Gallimard, Paris.

8 See the remarkable study by Albert-Llorca, M. (1991) *L'ordre des choses. Lés récits d'origine des animaux et des plantes en Europe*. Editions du Comit des Travaux Historiques et Scientifiques, Paris.
9 Translated from the Italian, Wittgenstein, L. (1975) *Note sul ramo d'oro di Frazer*. Adelphi Edizioni, Milan, p. 35.
10 I will cite just one source: Shiva, V. (1993) *Monocultures of the Mind. Perspectives on Biodiversity and Biotechnology*. Zed Books, London.
11 . . . e l'ala, i cieli, i boschi, la canzone: /i boschi antichi, ove una foglia casca,/ muta, per ogni battito di cuore. (G. Pascoli, Tristezze)
12 Geertz, C. (1973) *The Interpretation of Culture*. Basic Book, New York. Geertz, C. (1983) *Local Knowledge. Further Essays in Interpretative Anthropology*. Basic Book, New York. As for an interesting Italian analysis on the concept of 'local' in the relationship between man and environment see La Cecla, F. (1992) *Mente locale*. Laterza, Milan. Cecla, F. (1988) *Perdersi. L'uomo senza ambiente*. Laterza, Milan.
13 . . . Lontani dai nostri occhi vivono i boschi /chiusi con antiche parole, rovine d'altri tempi/ vivono dove non siamo più noi.
14 The Museo del Bosco (museum of the wood), opened to the public in 1993, displays materials of research carried out from 1989 to 1991. The method is based on oral sources, but there is also use of photography, archives, collections of objects and cartography. For a presentation of the museum project, see the catalogue edited by Molteni, G. (1993) *Il Museo del Bosco, Orgia*. Protagon editori toscani, Siena. We also refer to V. Lapiccirella-Zingari, doctoral dissertation, 1996, *Il bosco della memoria, la costruzione sociale dello spazio rurale nella memoria degli abitanti di una comunità toscana*. Siena University. This is consultable in the museum.
15 There is a rich literature on classic sharecropping, both as a system of agrarian management and as a form of life and culture. As regards Tuscan sharecropping see Clemente, P. (1988) *Il mondo a metà. Sondaggi antropologici sulla mezzadria classica*, Annali dell'Istituto Alcide Cervi, 9/1987. Società editrice Il Mulino, Imola.
16 On the ambiguity of the forest and its symbolic density, see Larrère, R. and Nougarède, O. (1993) *Des hommes et des forêts*. éd. Découvertes Gallimards, Eureux. Harrison, R. (1992) *Forêts, Essai sur l'imaginaire occidental*. éd. Flammarion, Paris. We also recommend the above mentioned authors' essays in Meiller, D. and Vannier, P. (eds) (1995) *La forêt, les savoirs et les citoyens, regards croisés sur les acteurs, les pratiques et les représentations*. éd. ANCR, Chalon sur Saône. As regards the wood as a place of imagery and the charcoal burner as the symbol of wood values, ranging from familiarity to extraneity, see Nesi, A. (1988) *Profumi d'altri tempi*, ed. Pantagruel, Pistoia; Clemente, P. (1996) In: Molteni, G.F. (ed.) *'Ottava vita' e dintorni, i carbonai dall'ottava rima al rock*, Tipografia dell'Amministrazione provinciale di Siena; Ferretti, R. (1977) *Fiabe e storie della Maremma nel fondo di tradizioni rurali della Maremma grossetana*, Roccastrada; Rosati, C. (1984) I carbonai, la lunga durata d'un mestiere. In: Rosati, C. *et al.* (ed.) (1984) *I mestieri del bosco, materiali per una documentazione*, Legato Antonini, Pistoia.
17 The themes of a sustainable development and the pattern of rural economies able to combine intensive use of resources, care for ecological equilibrium and even increase in biodiversity are today taken seriously by the international scientific community. As regards Europe, see Salbitano, F. (1988) *Human Influence on Forest Ecosystems Development in Europe*. Pitagora editrice, Bologna. Also see the essays by

Quezel P., Barbero, M., Bonin, G. and Loisel, R. (1992) Pratiques agricoles et couvert forestier en région méditerranéenne. in Dubost, M. (ed.) (1992) *Montagnes et forêts méditerranéennes*, éd. Icalpe, Savoie Technolac; and the recent issues of the FAO Journal (1996) *Communautées forestières dépendant de la forêt*, Vol. 47/186, No. 3, Unasylva, FAO Journal (1995) *Gestion de ressources forestières communautaires*, Vol. 46/180, No. 1, Unasylva. Also fundamental for a sustainable management of the resources and an effective fulfilment of the projects of sustainable development are the IK themes (indigenous knowledges) stressed by anthropology. See the essential work edited by Brokensha, D., Warren, D.M., and Werner, O. (1980) *Indigenous Knowledge Systems and Development*. University Press of America, Lanham, New York. See also Warren, D.M., Slikkerveer, L.J. and Brokensha, D. (1995) *The Cultural Dimension of Development. Indigenous Knowledge Systems*. Intermediate Technology Publications, London.

18 As for the feast as an important moment of community sharing and mutual acknowledgment of the different groups of rural society in the woods, see Jolas, T. and Zonabend, F. (1973) *Gens du finage, gens du bois*. Les Annales ESC, Paris.

19 Many French essays deal with the transformation of rural space perceived and expressed as loss or end of order. See Jolas, T., Pingaud, M.C., Verdier, Y. and Zonabend, F. (1990) *Une campagne voisine, Minot, village bourguignon*. éd. de la Maison de sciences de l'homme, Paris; and Fabre, D. (1986) *Le sauvage en personne*, Terrain No. 6. Carnets du patrimoine ethnologique, de la Maison du Patrimoine Ethnologique, Paris (1986), pp. 6–18. The recent book Vaisenat, C. and Notteghem, P. (eds), *Paysage au pluriel, pour une approche ethnologique des paysages*, éditions de la Maison de sciences de l'homme, Paris (1995), contains the results of various researches carried out in France. As for forest landscapes see Corvo, A., Arnould, P. and Hotyat, M. (1997) *La forêt, perceptions et représentations*. Editions L'Harmattan, Paris. As regards the present transformation of practices of use of forest resources, such as hunting and gathering, see Bromberger, C. and Lenclud, G. (1982) *La chasse et la cueillette aujourd'hui. Un terrain de recherche anthropologique?* Etudes Rurales, No. 87–88, Edition de l'Ecole des Hautes Etudes en Sciences Sociales, luglio–dicembre 1983, Paris, pp. 7–35.

20 Nei più nascosti recinti dell'acqua il ramo / il vero ramo arriva protendendosi / sempre più verde del suo non-arrivare / Proteggi l'astuzia soave dell'erba / dall'ordine attento delle biade / delle loro verdissime spade / in cui si taglia e s'intaglia l'estate / Voi molli onnipresenze / e folla di sorprese / fittissimamente conversate / sempre crescenti intese . . . Personal communication from Andrea Zanzotto to N. Breda. Now in *Il Gallo Silvestre*, n. 12, 2000, pp. 47–48, Editrice 'I Mori', Siena, Italy. In the same review there is the Italian version of the present chapter: *Breda, N. Tèra fréda, vergine, no se pal lavorarla*, 'Il Gallo Silvestre' n. 12, 2000, pp. 49–50, Editrice 'I Mori', Siena, Italy.

21 In northern Italy there are many areas called 'palù'. In this article we deal with the Palù del Quartier del Piave, in the Province of Treviso, investigated in Breda, N. (1993) the dissertation by *Palù. Una ricerca di etnoscienza sui palù del Quartier del Piave*. Università di Venezia. For ethnosciences' studies about the *padù*, see Breda, N. (1995) *Tassonomie popolari, ed osservazioni demologiche su specie vegetali di area trevigiana (Mosnigo di Moriago)*, in *Saggi dialettologici in area italo-romanza*, a cura di G.B. Pellegrini (ed.) CNR, Padova; and Breda, N. (1995) *Tassonomie botaniche popolari nei palù del Quartier del Piave* in 'Quaderni di semantica' a. XVI, n. 1 (1995): for a reflection on the passage from ethnoscience to an 'anthropology of knowledge', see Breda, N. (1997) *Figure sociali dei saperi naturali. Un percorso tra testi e rappresentazioni dall'etnoscienza all'antropologia*

dei saperi in '*La Ricerca Folklorica*' no. 36, ott. 1997, pp. 111–131, Grafo, Brescia, Italy. For a historical and agronomical approach to the palù see Gruppo 'Romìt', Ghizzo, L. (1997) *Il paesaggio agrario della Pianura Centro Occidentale.* Grafiche Antiga, Crocetta del Montello, Treviso, Italy.

22 The palù are a wide area which opposes the typical management of agriculture. This territory is a grassland intersected by a network of trees forming the typical structure of bocages.

23 Simmel, G. (1923) *Soziologie. Untersuchungen uber die Formen der Vergesellschaftung,* Duncker-Humblot, Munchen-Leipzig. A discussion of this theme is in Pozzi, Enrico (1993) *Lo straniero interno,* ed. Ponte alla Grazie, Florence.

24 For the concept of 'eco-cosmology' see Croll, E. and Parkin, D. (eds) (1996) *Bush Base: Forest farm. Culture, Environment and Development.* London, Routledge (essays by E. Croll, D. Parkin, T. Ingold, M. Leach, G. Persoon, etc.).

25 Farmers, in fact, see the palù as a source of 'richness', because they hardly need any labour or external intervention, but they supply wood, hay, products of gathering, etc.

26 Many tales on the origin of the palù have been reported. They tell of a large plain surrounded by mountains and covered by water. In these accounts various human interventions and various reasons (where myth and history are interwoven) have allowed the reflux of water and the emergence of the plain, except for the palù, which have ever since remained wetlands.

27 As for the much debated issue of a motorway project scheduled on the palù of San Vendemiano, San Fior, Godega S.U., see Breda, N. (2000) *Palù: paesaggi veneti e culture del nordest, tra conservazioni e devastazioni,* in C. Papa (eds) *Antropologia ambientale,* in 'La Ricerca Folklorica', n. 41, aprile 2000, pp. 13–20, Grafo, Brescia, Italy. See World Wildlife Fund, sez. Villorba, Treviso, *Un intero sistema di risorgive alberato sepolto dall'autostrada. Studio di Impatto ambientale dell'Autostrada A28 S. Vendemiano-PN,* Villorba 1990; WWF sez. Villorba, *Monitoraggio ambientale A28 Conegliano-PN, Lotti 28 e 2: 73 corsi d'acqua in gran parte di risorgiva cementificati o interrati,* Villorba 1994; WWF sez. di Villorba, *Un patrimonio naturale in un patrimonio storico-culturale: i palù tra Livenza e Monticano,* Villorba 1998; WWF sez. di Villorba, *L'ambiente inesistente ed i poteri rampanti,* Villorba 1997.

28 A. Zanzotto has devoted some verses to the palù, e.g., in his collection *Verso i palù,* which contains the extract we have quoted. The palù constitute the environmental background in his recent collection: *Meteo,* Donzelli (1996).

29 See Bateson, M.C. (1989) *Dove gli angeli esitano.* Adelphi, Milan, p. 301.

Prehistoric Cultures and the Development of Woodlands

18

H. Küster

Universität Hannover, Institut für Geobotanik, Nienburger Straße 17, D-30167 Hannover, Germany

Prehistoric cultures are called 'pre-historic' because written historic sources about their history were not produced in prehistoric times. It is therefore necessary to quote other sources and evidence to reconstruct the history of prehistoric cultures and their environment. Pollen diagrams give very detailed evidence about the development of the environment in both prehistoric and historic times.

1 Principles of pollen analysis

Pollen grains are produced by flowering plants; in these grains the male genetic information is transported from male to female flowers. In many cases insects such as bees transport the pollen grains from one flower to another. In other cases pollen grains are distributed by the wind. If the wind is the vector for pollen transport, it is very unlikely that pollen grains are deposited just on a female flower after transport. Plants distributing their pollen grains by the wind produce a lot of pollen grains; most of them are deposited on the earth's surface as 'pollen rain'. If the pollen rain is deposited on the surface of a bog the pollen grains are not destroyed but included in the accumulating peat. If the pollen grains are deposited on a lake surface, they sink to the lake floor and are later included in the accumulating lake sediment. In these sediments pollen grains can be preserved for millennia.

Sediment can be extracted from a bog or lake floor by coring a total section through all layers. The section is cut into samples which are time slices: each

 Methods and Approaches in Forest History (eds M. Agnoletti and S. Anderson)

slice contains the sediments and the pollen grains which have been deposited in a specific period of time. The pollen grains can be determined and counted under a microscope, and the sediments can be dated by carbon-14 measurements. After completing a pollen analysis it is known which plant taxa grew in the vicinity of the bog or lake. It is possible to identify cereal pollen grains, which enables one to recognize when farming was started by prehistoric cultures, which is an indicator of active management of woodlands and the landscape.

2 Impact of early agriculture on woodlands

The first prehistoric agricultural groups in many parts of Europe were Neolithic cultures which developed from the sixth millennium BC. Agriculture was based on the cultivation of crops which were not indigenous to Europe. Crops were first cultivated in western Asia and distributed to Europe afterwards. The landscape had to be changed completely by prehistoric farmers to allow for the introduction of crop cultivation. After the change of the environment Europe became an ideal area for agriculture, where farming was less likely to be plagued by natural catastrophes than other parts of the world (Küster, 1995).

Before the establishment of agriculture, Europe was almost completely wooded. Oak was predominant in many European woodlands – as can be detected from the pollen diagrams. Woodlands had to be cleared to allow for the growth of plants such as cereals and pulses which derived from western Asian steppe regions. The clearance of woodland was not practised in the same way as in the tropical woodlands of the present day. Wood was by no means wasted in Europe. The trunks were used as timber to construct houses. From archaeological excavations it is evident that the first farmers in Central Europe constructed long houses with a length of 30–35 m at maximum. These houses could only be stable if pieces of timber 30–35 m long were used as ridges for the roofs. It is striking that oak trees are exactly this length, indicating that it is very likely that oak trees were cut in the vicinity of the first settlements and that the whole trunks were used to construct the long houses. Wood was also necessary as fuel for many purposes such as heating, baking and cooking. On the cleared area near to the settlement crops could be grown. The Neolithic farmers also brought domesticated animals to Europe. Cattle grazed the woodlands surrounding the clearing, as open pastures were not yet in existence. Cattle browsed not only grass and herbs growing in the woodlands but also twigs and young shoots of the trees. Pigs fed on acorns which otherwise could have germinated and grown into trees. The browsed woodlands became more and more open through time. As meadows did not exist in prehistoric times grass and herbs could not be cut to make hay. Instead of this, twigs were cut by prehistoric farmers which were dried, preserved and fed to the cattle in wintertime. Elm and ash trees were particularly used, but trees of other species were also pollarded to collect leaf hay.

All these activities can be detected in a pollen diagram. Cereal pollen grains give evidence that agriculture was practised. Where the first pollen grains are found, the percentages of some tree species became smaller in the same layers of the peat or lake marl. In particular the pollen percentages of elm, which does not survive intensive and repeated pollarding, became smaller as a consequence of human activities; the so-called 'elm decline' is very characteristic in many pollen diagrams, especially those deriving from north-western Europe (Troels-Smith, 1960; Peglar, 1993). It is generally acknowledged that in connection with the spread of agricultural settlements and activities woodlands were cleared, reducing the total woodland area.

From archaeological excavations we know that prehistoric settlements did not exist longer than a few decades in the same places. After this they were abandoned, and new settlements were founded in other areas (Waterbolk, 1982; Kossack, 1997; Lüning *et al.*, 1997). In some cases the total history of foundations, abandonments and dislocations of a settlement through time can be traced by excavations (Hvass, 1982). It has often been assumed that the crop yields became smaller with time so that the farmers were forced to leave their fields and settlements after some decades; however, it is more likely that shortages of timber were the reason for the relocation of settlements. After the foundation of a settlement and during its existence, timber was not just available at its locality. As the houses were timbered and as open fires were burning inside, occasionally houses burned down. After a fire, timber was not available at the settlement location. Outside the range of the settlements, primeval woodlands still occurred. Timber could be transported from there to the settlement location. As the transport of 30 m long oak trunks was, of course, very complicated it was often easier to relocate the settlement to a place where timber was still available (Küster, 1998).

In the abandoned settlement areas woodlands could regenerate. Pollen analytical results show that only in some coherent layers of the profiles were cereal pollen grains deposited. After this more birch pollen grains were deposited, indicating that a secondary succession of woodlands had started on the abandoned cereal fields and the settlement area. Under the birches other tree species could grow up. From the pollen analytical results it is evident that not only oak became more frequent again, but also other tree species, such as beech in many parts of Central Europe (Küster, 1996, 1997), hornbeam in eastern Central Europe (Ralska-Jasiewicz, 1964) and spruce in Switzerland (Markgraf, 1970, 1972). As a result of secondary succession the primeval woodlands did not develop again, but the succession led to the formation of another type of woodland.

The prehistoric period lasted for several millennia in many parts of Europe. During these millennia settlements were founded and abandoned again many times, so that in the areas which were favourable for agriculture nearly every patch of woodland was cleared at least once during this time span. Over the course of millennia, therefore, each patch of woodland in an area was transformed. Instead of primeval woodlands, new types of woodland became established. The spread of beech woodlands in Central Europe lasted several thousands of years; it started in the fifth millennium BC when the first prehistoric settlements were founded in this area. Beech was spreading until the beginning of the Middle Ages around 500 AD. In this period beech reached the outermost boundaries of its growing area in south-east England, southern Scandinavia and the Polish coast of the Baltic Sea.

3 Impact of early ore mining on woodlands

In the Iron Age, in the first millennium BC, men started to use woodlands in another way. Mining of ores such as copper and iron became important as tools and implements were constructed from ores. Ores had to be smelted in small furnaces to extract them from rocks and to form implements. The furnaces were heated with firewood. Woodlands were used continuously in the areas around the ore-smelting places, and secondary successions of woodlands were not possible in these areas. The expansion of beech came to an end in these areas; this is visible in the region north of the Alps (Küster and Rehfuess, 1997) and in the hilly landscapes in western Germany such as the Siegerland (Pott, 1985, 1986). As the management of the woodlands near to the furnaces became intense, new beech shoots could not grow up from tree stumps. Tree species forming new shoots from stumps even after intensive management could spread. From pollen analytical results it is seen that hornbeam, birch and other trees became more frequent (Pott, 1981).

4 Settlements and woodlands in historic times

From the beginning of historic times onwards, in some regions since Roman times and in other regions beginning in the Middle Ages, settlements became more and more stable. Only stable settlements could be included in empires which expanded throughout Europe in these ages. The transformation of unstable to stable settlements caused a marked change in woodland development which is evident from pollen analytical results.

In each layer of a sediment section cereal pollen grains occur indicating that agriculture was practised permanently. Beech did not continue to spread, as mentioned above. In the Early Middle Ages, in the middle of the first millennium AD, beech had reached the outermost boundaries of its actual growing area in Europe. Secondary successions of woodlands after the abandonment of a settlement only rarely occurred so that beech could not continue to spread. Fields were used continuously, and the same patches of woodland were used to get timber and firewood. In these intensively managed woodlands beech became rarer, and hornbeam and birch became more frequent as in the woodlands in the ore-smelting areas in the millennium before.

Local shortages of timber were the consequence of the transformation of the settlement strategy as big trunks were not always available near to the stable settlements when they were needed. In the developing empires it was necessary to establish timber transport and trade which, in fact, was a precondition for the existence of stable settlements. To organize timber transport and trade, writing was necessary; therefore we can use written sources from the Middle Ages and modern times to reconstruct woodland history today.

5 Final remarks

It was not necessary to use writing to organize the economy in prehistoric settlements, thus we do not have written sources from prehistoric periods. Pollen analysis in connection with archaeological results have to be used to reconstruct the early development of woodlands and the environment of man in total. It is obvious that woodlands had been transformed by man intensively before the Middle Ages. Man did not only actively influence the development of woodlands but also through allowing secondary successions of woodlands. The spread of beech and other tree species was furthered by man. On the other hand, the decline of oak in places, where later beech became more frequent, the declines of elm in the Neolithic, and the declines of beech in the ore-smelting areas were also caused by human activities.

When the first written sources were produced woodlands were already intensively transformed to a state which was far removed from the primeval state, so that they cannot be regarded as natural. To detect the primeval, natural state of woodlands pollen analysis is always necessary; the analysis of written sources and the analysis of the actual vegetation alone can never reconstruct the composition of primeval woodlands.

References

Hvass, S. (1982) Ländliche Siedlungen der Kaiser- und Völkerwanderungszeit in Dänemark. *Offa* 39, 189–195.

Kossack, G. (1997) *Dörfer im nördlichen Germanien, vornehmlich aus der römischen Kaiserzeit, Lage, Ortsplan, Betriebsgefüge und Gemeinschaftsform*, Bayerische Akademie der Wissenschaften, Philosophisch-Historische Klasse, Abhandlungen. *Neue Folge* 112. Beck, Munich.

Küster, H. (1995) *Geschichte der Landschaft in Mitteleuropa*. Beck, Munich.

Küster, H. (1996) Auswirkungen von Klimaschwankungen und menschlicher Landschaftsnutzung auf die Arealverschiebung von Pflanzen und die Ausbildung mitteleuropäischer Wälder. *Forstwissenschaftliches Centralblatt* 115, 301–320.

Küster, H. (1997) The role of farming in the postglacial expansion of beech and hornbeam in the oak woodlands of central Europe. *The Holocene* 7, 239–242.

Küster, H. (1998) Auswirkungen prähistorischen Siedelns auf die Geschichte der Wälder. *Regensburger Beiträge zur Prähistorischen Archäologie* 5, 23–40.

Küster, H. and Rehfuess, K.E. (1997) Pb and Cd concentrations in a southern Bavarian bog profile and the history of vegetation as recorded by pollen analysis. *Water, Air, and Soil Pollution* 100, 379–386.

Lüning, J., Jockenhövel, A., Bender, H. and Capelle, T. (1997) *Deutsche Agrargeschichte.* Vor- und Frühgeschichte, Ulmer, Stuttgart.

Markgraf, V. (1970) Palaeohistory of the spruce in Switzerland. *Nature* 228, 249–251.

Markgraf, V. (1972) Die Ausbreitungsgeschichte der Fichte (*Picea abies* H. Karst.) in der Schweiz. *Berichte der Deutschen Botanischen Gesellschaft* 85, 165–172.

Peglar, S. (1993) The mid-Holocene *Ulmus* decline at Diss Mere, Norfolk, UK: a year-by-year pollen stratigraphy from annual laminations. *The Holocene* 3, 1–13.

Pott, R. (1981) Der Einfluß der Niederholzwirtschaft auf die Physiognomie und die floristisch-soziologische Struktur von Kalkbuchenwäldern. *Tuexenia* 1, 233–242.

Pott, R. (1985) *Vegetationsgeschichtliche und pflanzensoziologische Untersuchungen zur Niederwaldwirtschaft in Westfalen.* Westfälisches Museum für Naturkunde, Münster.

Pott, R. (1986) Der pollenanalytische Nachweis extensiver Waldbewirtschaftung in den Haubergen des Siegerlandes [Palynological evidence of extensive woodland management in turn with agriculture in the area of the Hauberge in Siegerland, Northrine-Westphalia, FRG]. In: Behre, K.E. (ed.) *Anthropogenic Indicators in Pollen Diagrams.* Balkema, Rotterdam, pp. 125–134.

Ralska-Jasiewicz, M. (1964) Correlation between the Holocene history of the *Carpinus betulus* and prehistoric settlement in north Poland. *Acta Societatis Botanicorum Poloniae* 33, 461–468.

Troels-Smith, J. (1960) Ivy, mistletoe and elm. Climate indicators, fodder plants. *Danmarks Geologiske Undersøgelse* 4. Reitzel, Copenhagen.

Waterbolk, H. T. (1982) Mobilität von Dorf, Ackerflur und Gräberfeld in Drenthe seit der Latènezeit. Archäologische Siedlungsforschungen auf der nordniederländischen Geest. *Offa* 39, 97–137.

19 Integration between Genetic and Archaeobotanical Data in a Study on the Evolutionary History of *Pinus halepensis* Mill. Populations in Southern Italy

S. Puglisi[1], G. Fiorentino[2], R. Lovreglio[3] and V. Leone[3]

[1]Istituto del Germoplasma, Consiglio Nazionale delle Ricerche, Via Amendola 165/A, I-70126 Bari, Italy; [2]ESA 5059 CNRS LAB PAAH, Institut de Botanique, 163 rue A. Broussonnet, 34095 F Montpellier Cedex 2, France; [3]Dipartimento di Produzione Vegetale, Università della Basilicata, Via Nazario Sauro 85, 85100 Potenza, Italy

The first results of multidisciplinary research on the evolutionary history of *Pinus halepensis* Mill. are presented here. For this purpose the genetic structures of nine natural populations – four Apulian, four Calabrian and one from Basilicata – were studied using isozyme markers. The obtained results were integrated with archaeobotanical analysis carried out on several archaeological sites concerning a chronological range of 30,000 years, showing the presence of *Pinus* cf. *halepensis* in Apulia only from 17,000 to 10,000 years BP. Since further palaeobotanical and genetic information is not available, at present we can only hypothesize on its anthropical introduction into the Apulian region, probably as a consequence of colonization events from the Aegean area in historical times. In this case the genetic similarity observed in several studies between Apulian and Greek populations – which are also characterized by the highest values of genetic diversity parameters, in comparison with populations from the remaining regions of the natural range – could not be a consequence of natural migratory events but of the provenance of propagation material used to establish the present Apulian stands.

1 Introduction

Aleppo pine (*Pinus halepensis* Mill.) is a circum-Mediterranean species (Fig. 19.1) characterized by high tolerance to drought and to adverse pedological conditions, and therefore particularly suitable to reforestation of arid zones. Moreover, the high adaptation of this species to Mediterranean bioclimates is confirmed by some features of its reproductive cycle – in particular the development of the female gametophyte (Francini, 1958) – and by production of serotinous cones which enable the settlement of abundant natural regeneration after fires (Saracino and Leone, 1993a,b; Saracino *et al.*, 1997). It is related to the morphologically similar species *Pinus brutia* Ten., whose main distribution area is restricted to the eastern Mediterranean basin, characterized by more regular trunk shape and greater cold tolerance. *P. brutia* was previously considered to be a subspecies of *P. halepensis*, but its species status was afterwards fully recognized on the basis of biochemical traits. Conkle *et al.* (1988), on the basis of results obtained by means of biochemical genetic markers, hypothesized that *P. halepensis* was derived from *P. brutia*-like progenitors, and that this differentiation and the westwards colonization of the Mediterranean basin starting from the centre of origin of this species complex were accompanied by a remarkable loss of genetic variation.

Both species were subjected to the FAO 4-bis Project 'International experiences upon provenances of *Pinus halepensis* and *Pinus brutia*', which started

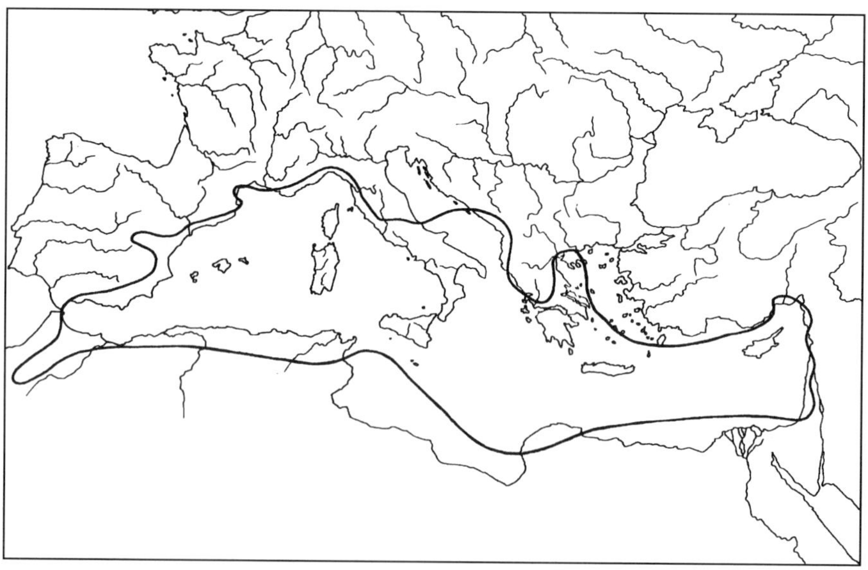

Fig. 19.1. Natural range of *Pinus halepensis* Mill. (after Mirov, 1967).

in the 1970s and gave rise to numerous studies on provenance variation of morphological, physiological and biochemical traits.

Despite its wide distribution, this species is considered of modest importance and few studies exist on its population genetic structure, made by means of biochemical genetic markers. The most important research (Schiller *et al.*, 1986) was carried out by means of isozyme analysis on 26 populations: eight Israeli and the remnant distributed all over the natural range. Four of the studied populations are Italian: two of them are Apulian, one Umbrian and one Ligurian; they were included in different groups on the basis of the surveyed parameters.

In Italy, the surface covered by Aleppo pine high forests covers about 20,000 ha, half of which are in Apulia (Magini, 1955; Bernetti, 1995).

The aim of this research is the study of the genetic structure of nine typical Aleppo pine populations, estimated by means of isozymes, and the comparison of genetic with archaeological data in order to reconstruct the evolutionary history of this species in southern Italy.

2 Materials and methods

The nine studied populations (Fig. 19.2) are reputed to be of natural origin and are located in four areas of southern Italy: Gargano (Apulia; two populations), Apulian Ionian coast (two populations), Calabrian Ionian belt (four populations), Basilicata (one population). In the first eight populations, sampling was carried out by collecting cones from ten trees per population, uniformly placed on the selected area; only in the ninth population were cones collected from 20 trees.

For the first eight populations isozyme analysis was performed on haploid megagametophytes and embryos of 12 seeds per mother tree by means of starch gel horizontal electrophoresis, in order to distinguish between male and female gametes in each embryo. Because of the small number of sampled mother trees, only paternal alleles were used for estimating allele frequencies, in order to obtain a more representative sample than the one constituted by embryo genotypes. On the basis of pollinic allele frequencies estimated at 13 loci of seven enzyme systems (LAP, GOT, PGM, 6-PGDH, MDH, GDH, SKDH), several parameters of genetic diversity (variation within populations) and genetic differentiation (variation between populations) were computed.

As regards the final population, because of the close proximity of some artificial stands and the consequent risk of pollen contamination, analysis was performed only on six megagametophytes per tree, in order to use only the genotypes of the 20 sampled mother trees for estimating allele frequencies at the above-mentioned loci and computing diversity and differentiation parameters (Puglisi *et al.*, 1999).

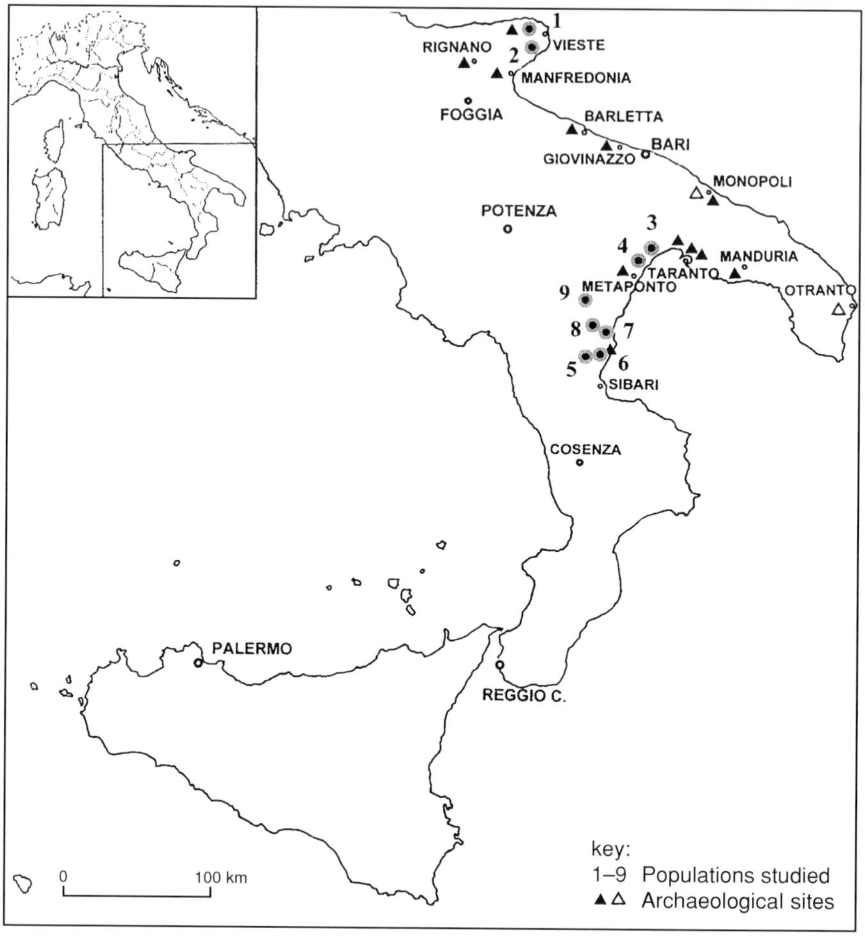

Fig. 19.2. Location of the studied populations (grey and black circles). 1: Coppa della Nuvola (Paschici, Foggia – Gargano, Apulia); 2: Monte Barone (Mattinata, Foggia – Gargano, Apulia); 3: Perronello (Riva dei Tessali, Taranto – Apulia); 4: Pineta della Regina (Marina di Ginosa, Taranto – Apulia); 5: Albidona 1 (Cosenza – Calabria); 6: Albidona 2 (Cosenza – Calabria); 7: Stràtolo (Valle del Ferro, Cosenza – Calabria); 8: Masseria Sorìa (Valle de Ferro, Cosenza – Calabria); 9: Tursi (Matera – Basilicata). Triangles mark the 14 archaeological sites subjected to archaeobotanical analysis; the white triangles correspond to the only two sites where carbonized fragments of *P. halepensis* wood, dating back to the Late Glacial Period, were present (see text).

The archaeobotanical analysis was carried out on 14 archaeological sites (Fig. 19.2) covering a chronological range of 30,000 years, by studying carbonized wood fragments under an optical light/dark field reflected light microscope, and in some cases with a scanning electron microscope, in each of the three traditional anatomical planes of section (transversal, tangential and radial), obtained by manual fracturing. The taxonomic identification of fragments was based on comparison with contemporary carbonized specimens present in collections and on wood anatomy atlases (Greguss, 1955; Jacquiot, 1955; Zucconi, 1959; Schweingruber, 1990). In particular, the anatomical characteristics in radial section, with pinoid pits in cross-fields, and the position of resin canals near growth ring boundaries in transverse section, were utilized for discriminating between Mediterranean *Pinus* species (*Pinus halepensis* Mill. in relation to *Pinus pinea* L. and *Pinus pinaster* Aiton). For *Pinus halepensis* Mill. and *Pinus brutia* Ten., these two species cannot be distinguished from one another on the basis of their wood anatomy. Despite these limitations, a greater taxonomical resolution is possible by means of anatomical characterization of charcoal fragments, in comparison with palynological analysis which only enables recognition of the genus *Pinus*.

3 Results and discussion

As regards genetic diversity parameters, on the whole, the Apulian populations are the most variable. But considering the most significant parameters – expected heterozygosity according to Hardy-Weinberg (Nei, 1975) and, especially, effective number of alleles (Gregorius, 1978) – populations no. 7 and no. 9 (from Calabria and Basilicata, respectively; Fig. 19.2) also place themselves on a level with the Apulian ones.

Other studies (Schiller *et al.*, 1986; Morgante *et al.*, 1998) – carried out by means of biochemical and molecular genetic markers – have emphasized the highest genetic diversity of populations from Gargano, which are second only to Greek ones in the whole range of this species. However, the values we obtained are considerably higher than the values obtained by other authors for populations close to ours (Schiller *et al.*, 1986). The genetic diversity values we obtained, however, do confirm results of former research (Grunwald *et al.*, 1986; Schiller *et al.*, 1986; Teisseire *et al.*, 1995; Agúndez *et al.*, 1997): genetic variation in this species as a whole is rather low, in comparison with variation estimated in most of the studied conifers (Hamrick *et al.*, 1981).

As regards the genetic differentiation parameters, the relative diversity between populations (G_{st}) (Nei, 1973, 1975) is 3%, similar to many other wide-ranged conifers, thus showing that the overwhelming majority of the measured genetic diversity is within populations. The obtained Nei's genetic distance values (Nei, 1972) clearly show the trend of the studied populations to group into three distinct clusters (Fig. 19.3):

1. Populations from Apulia (nos 1, 2, 3 and 4) and Basilicata (no. 9);
2. Calabrian populations from Valle del Ferro (nos 7 and 8);
3. Calabrian populations from Albidona (nos 5 and 6).

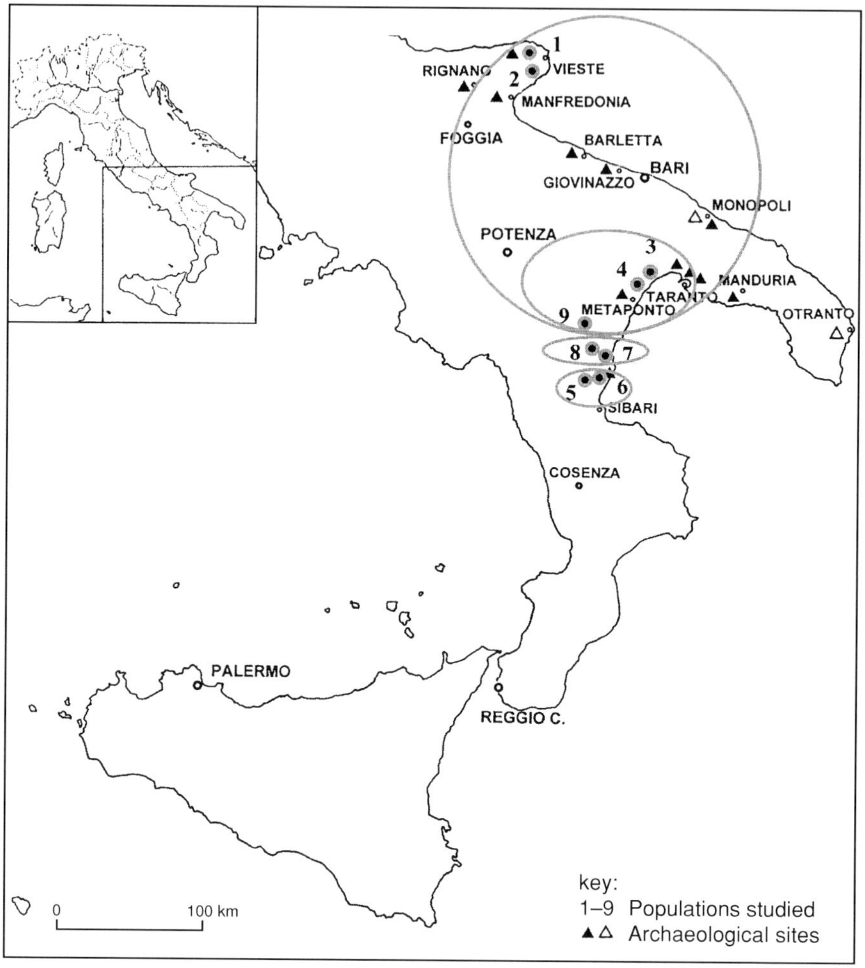

Fig. 19.3 Population clusters resulting from values of Nei's genetic distance between populations.

In particular, the two populations from Albidona (nos 5 and 6) are sharply differentiated from the rest, including the other two Calabrian populations (Valle del Ferro, nos 7 and 8) in spite of the short geographic distance which separates them. It is also worthy of note that the population from Basilicata (no. 9) is genetically very similar to the populations nos 3 and 4 (from the Apulian Ionian coast), which enables us to hypothesize a common origin. The trend to differentiation of the populations from Gargano (nos 1 and 2) is also evident. In Fig. 19.4 the population clusters are shown at a higher level of genetic distance.

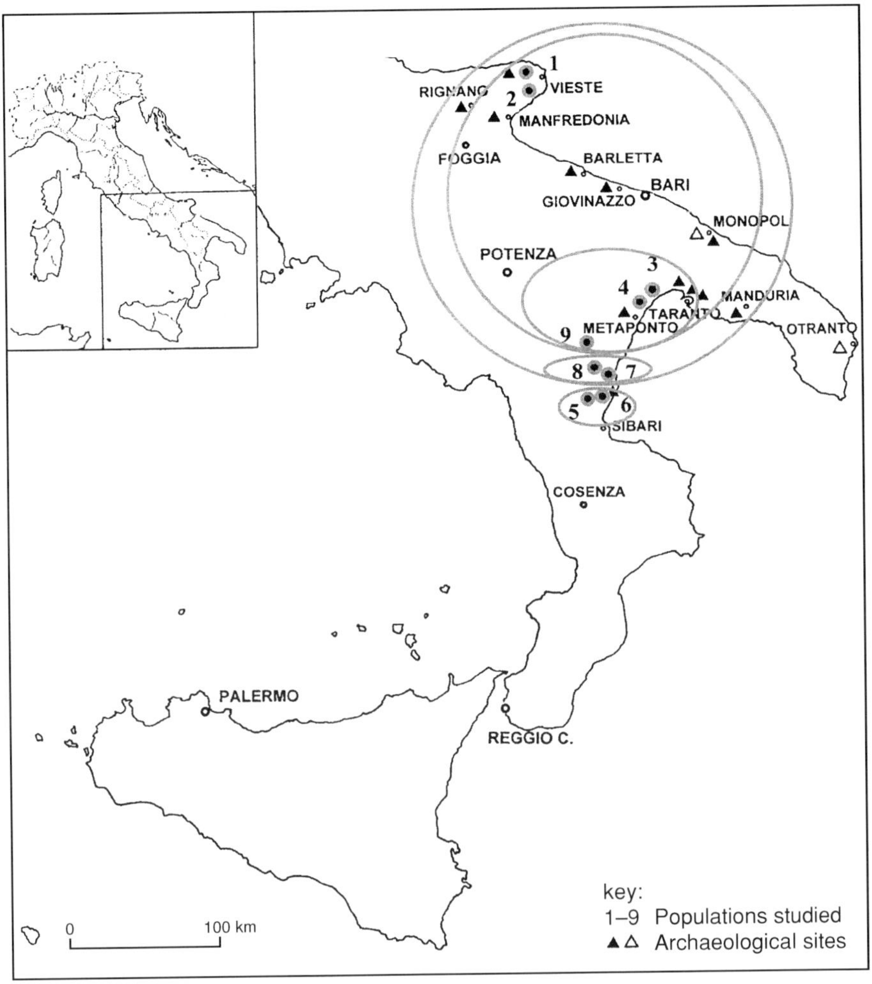

Fig. 19.4. Population clusters shown at a higher genetic distance level than in Fig. 19.3.

In connection with their differentiation, the Calabrian populations are also distinguishable in their lower genetic variation. Since these populations are accepted as spontaneous, it can be hypothesized that their differentiation could result from the crossing in past times of one or more *genetic bottlenecks* which considerably reduced variation, probably because of wildfires occurring in young stands before the first seed production. The *bottleneck* crossing could have occurred in the course of a hypothetical migration and colonization of the Calabrian Ionian belt from the Apulian coast, since Aleppo pine is generally considered autochthonous in Apulia but not in Calabria (Francini, 1953; Bernetti, 1995). The population from Basilicata could also have

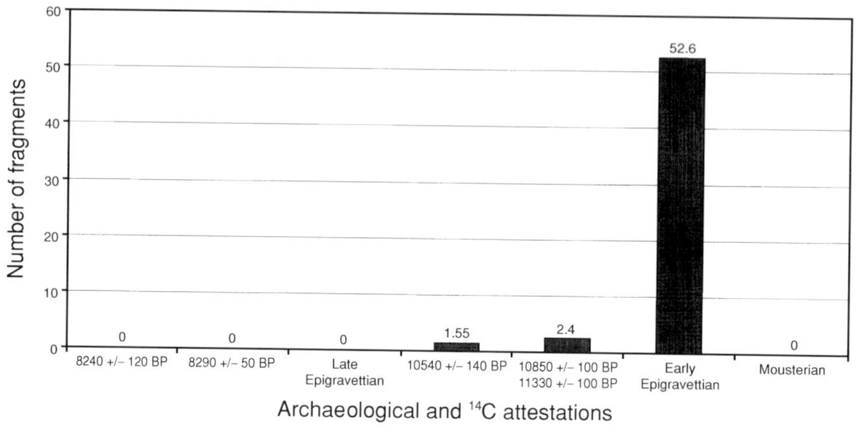

Fig. 19.5. Histogram of per cent fluctuations of *Pinus* cfr. *halepensis* Mill. fragments in the sequence of Le Mura Cave (Monopoli).

originated from a migratory movement which, starting from the Apulian Ionian coast, could have followed the coast of Basilicata and then ascended the valleys of the Agri and Sinni rivers, but in this case such a movement would not have been accompanied by a loss of genetic variation.

Nevertheless, the available archaeobotanical data do not seem to support the hypothesis of a natural origin of present Apulian populations (Fig. 19.2). Studies executed in Apulia only reveal the presence of carbonized fragments of *P. halepensis* wood, dating back to the Late Glacial Period (17,000–10,000 years BP), in Monopoli, on the Adriatic coast close to Bari (Fig. 19.5; Fiorentino, 1998a), and in Grotta Romanelli, near Otranto (Salentine Peninsula; Follieri, 1968). On the contrary, remains of this species are absent in the archaeological places studied in Vieste (Gargano; Fiorentino, 1995a) and in Manduria (eastern Ionian belt near Taranto; Fiorentino, 1995b), dating back to the Early Neolithic Age (about 7000 years BP) and to the Late Mesolithic/Early-Middle Neolithic Age (about 7200–6500 years BP), respectively. They are also absent in several places dating back to the Bronze Age (fourth to third millennium BP) situated along the Apulian Adriatic belt (Fiorentino, 1995c, 1998b; Fiorentino and Radina, 1998), including a place in Monopoli (Fiorentino, 1995d) where this species was previously present, and in three places situated along the Ionian belt near Taranto (at present covered by wide *P. halepensis* stands used in our sampling) dating back to the Bronze Age too (Fiorentino, 1999a, b). Moreover, a Calabrian Bronze Age place, situated near two of the populations we analysed (nos 5 and 6), revealed the absence of any Aleppo pine remains (Vallino and Ventura, 1984).

These data could represent a further, indirect confirmation of the hypothesis by Pignatti (1979) about the anthropic origin of the present populations of this species in Apulia: 'Although this statement seems to be rather iconoclastic, I would be of the opinion that also in these zones [Gargano and southern Apulia] *Pinus halepensis* is extra-zonal and in all probability introduced by man' (Pignatti, 1979).

As a consequence, we cannot exclude that the Calabrian populations too have an anthropic origin, or that the above-mentioned migration could have occurred in a more recent period. The genetic variation of the Apulian populations, higher than in the other Italian and Mediterranean populations, could originate from Greek provenance of the propagation material introduced by Greek colonizers present in southern Italy in historical times (*Magna Grecia*), since Greek populations show the highest variation values (Loukas *et al.*, 1983; Schiller *et al.*, 1986; Morgante *et al.*, 1998).

Acknowledgements

The authors wish to thank Mrs M. Attolico for her excellent technical assistance. This study was partly supported by the FIREGENE AIR3-Contract: CT93-0803.

References

Agúndez, D., Degen, B., von Wuehlisch, G. and Alia, R. (1997) Genetic variation of Aleppo pine *(Pinus halepensis* Mill.) in Spain. *Forest Genetics* 4, 201–208.

Bernetti, G. (1995) *Selvicoltura Speciale*. UTET, Torino.

Conkle, M.T., Schiller, G. and Grunwald, C. (1988) Electrophoretic analysis of diversity and phylogeny of *Pinus brutia* and closely related taxa. *Systematic Botany* 13, 411–424.

Fiorentino, G. (1995a) New perspectives in anthracological analysis. Palaeocological and technological implications of charcoals found in the Neolithic flintmine at La Defensola (Vieste, Apulia, Italy). *Quaternaria Nova* 5, 99–128.

Fiorentino, G. (1995b) Analisi dei macroresti vegetali [Analysis of botanical macroremains]. In: Gorgoglione, M.A., Di Lernia, S. and Fiorentino, G. (eds) *L'insediamento preistorico di Terragne* (*Manduria–Taranto*) [*The prehistoric site of Terragne*]. Regione Puglia – Centro Regionale Servizi Educativi e Culturali TA/55, Manduria, pp. 171–183.

Fiorentino, G. (1995c) Madonna del Petto (Barletta – Bari): analisi antracologica e caratteristiche dell'immagazzinamento dei carporesti [anthrocological analysis and storage processing features of botanical macroremains]. *Taras – Rivista di Archeologia*, 15. pp. 217–231.

Fiorentino, G. (1995d) Primi dati archeobotanici dall'insediamento dell'Età del Bronzo di Monopoli – Piazza Palmieri [First archaeobotanical data from the Bronze Age site of Monopoli – Palmieri Square]. *Taras – Rivista di Archeologia* 15. pp. 335–373.

Fiorentino, G. (1998a) L'exploitation du milieu par l'Homme Paléolithique à l'Age du Bronze en Italie sud-orientale (Pouilles): données anthracologiques et carpologiques. Thèse de Doctorat, Université de Montpellier II, 402pp.

Fiorentino, G. (1998b) Le risorse vegetali [Vegetal resources]. In: Radina, F. and Cinquepalmi, A. (eds) *Documenti dell'età del Bronzo. Ricerche lungo il versante adriatico pugliese* [*Archaeological data from Bronze Age. Researches along the Apulian Adriatic coastline*]. Schena Editore, Fasano, pp. 211–221.

Fiorentino, G. (1999a) Carratteristiche della vegetazione e abitudini alimentari durante la Preistoria [Vegetation features and nourishment in Prehistory]. In: Mastronuzzi, G. and Marzo, P. (eds) *Le isole Cheradi tra natura, laggenda e storia* [*Charad: Isles among nature, legend and history*]. Stampe Sud, Mottola (Taranto), pp. 69–78.

Fiorentino, G. (1999b) Paleo-ambiente e paleo-economia nell'arco jonico [Palaeoenvironment and palaeoeconomy along the Ionian coast]. In: *Atti del Seminario di Studi: Torre Castelluccia. Un insediamento dell'Età del Bronzo a Pulsano* [*Proceedings of the Workshop: Castelluccia Tower. A Bronze Age site in Pulsano*] (in press).

Fiorentino, G. and Radina, F. (1998) New data about food production and environment during Bronze age in South-Eastern Italy. In: *Proceedings of the 7th Meeting of the International Committee of Anthropology of Food: the Road of Food Habits in the Mediterranean area. Rivista di Antropologia* 76 (suppl.), 57–64.

Follieri, M. (1968) Determinazioni xilotomiche dei carboni del Paleolitico superiore di Grotta Romanelli in Terra d'Otranto [Xylotomic analysis of charcoals from the Upper Palaeolithic site of Romanelli cave in Otrantoland]. *Quaternaria* 10, 125–135.

Francini, E. (1953) Il pino d'Aleppo in Puglia [Aleppo pine in Apulia]. *Annali della Facoltà di Agraria dell'Università di Bari* 8, 309–416.

Francini, E. (1958) Ecologia comparata di *Pinus halepensis* Mill., *Pinus pinaster* Sol. e *Pinus pinea* L. sulla base del gametofito femminile [Comparative ecology of *Pinus halepensis* Mill., *Pinus pinaster* Sol. and *Pinus pinea* L. based on the female gametophyte]. *Annali dell'Academia Italiana di Scienze Forestali* 7, 107–172.

Gregorius, H.R. (1978) The concept of genetic diversity and its formal relationship to heterozygosity and genetic distance. *Mathematical Biosciences* 41, 253–271.

Greguss, P. (1955) *Identification of Living Gymnosperms on the Basis of Xilotomy.* Akadèmiai Kiadò, Budapest.

Grunwald, C., Schiller, G. and Conkle, M.T. (1986) Isozyme variation among native stands and plantations of Aleppo pine in Israel. *Israel Journal of Botany* 35, 161–174.

Hamrick, J.L., Mitton, J.B. and Linhart, Y.B. (1981) Levels of genetic variation in trees: influence of life history characteristics. In: Conkle, M.T. (ed.) *Proceeding of the Symposium on Isozymes of North American Forest Trees and Forest Insects.* General Technical Report PSW-48. USDA Forest Service, PSW Forest and Range Experiment Station, Berkeley, pp. 35–41.

Jacquiot, C. (1955) *Atlas d'anatomie des bois des conifères.* Centre Technique du Bois, Paris.

Loukas, M., Vergini, Y. and Krimbas, C.B. (1983) Isozyme variation and heterozygosity in *Pinus halepensis* Mill. *Biochemical Genetics* 21, 497–509.

Magini, E. (1955) Pinete di pino d'Aleppo [Aleppo pine forests]. *Atti Congresso Nazionale di Selvicoltura* 1, 49–68.

Mirov, N.T. (1967) *The Genus Pinus.* Ronald Press Company, New York.

Morgante, M., Felice, N. and Vendramin, G.G. (1998) Analysis of hypervariable chloroplast microsatellites in *Pinus halepensis* reveals a dramatic genetic bottleneck. In:

Karp, A., Isaac, P.G. and Ingram, D.S. (eds) *Molecular Tools for Screening Biodiversity.* Chapman & Hall, London, pp. 407–412.

Nei, M. (1972) Genetic distance between populations. *American Naturalist* 106, 283–292.

Nei, M. (1973) Analysis of gene diversity in subdivided populations. *Proceedings of the National Academy of Sciences of the USA* 70, 3321–3323.

Nei, M. (1975) *Molecular Population Genetics and Evolution.* North Holland Publishers, Amsterdam.

Pignatti, S. (1979) I piani di vegetazione in Italia [Vertical zonation of vegetation in Italy]. *Giornale Botanico Italiano* 113, 411–428.

Puglisi, S., Lovreglio, R., Cifarelli, S. and Leone, V. (1999) Allozyme variation in five populations of *Pinus halepensis* Mill. in southern Italy. *Forest Genetics* 6, 241–246.

Saracino, A. and Leone, V. (1993a) Ecological features and dynamics of seed dispersal after fire in a *Pinus halepensis* Miller forest. In: Trabaud, L. and Prodon, R. (eds) *Fire in Mediterranean Ecosystems.* Banyuls-sur-Mer, France, 21–25 September, 1992. Commission of the European Communities, Ecosystems Research Report No. 5. Guyot, Bruxelles, pp. 151–159.

Saracino, A. and Leone, V. (1993b) Natural regeneration 2 and 4 years after fire of *Pinus halepensis* Miller in dunal environment. In: Trabaud, L. and Prodon, R. (eds) *Fire in Mediterranean Ecosystems.* Banyuls-sur-Mer, France, 21–25 September, 1992. Commission of the European Communities, Ecosystems Research Report No. 5. Guyot, Bruxelles, pp. 141–150.

Saracino, A., Pacella, R., Leone, V. and Borghetti, M. (1997) Seed dispersal and changing seed characteristics in a *Pinus halepensis* Mill. forest after fire. *Plant Ecology* 130, 13–19.

Schiller, G., Conkle, M.T. and Grunwald, C. (1986) Local differentiation among Mediterranean populations of Aleppo pine in their isoenzymes. *Silvae Genetica* 35, 11–19.

Schweingruber, F.H. (1990) *Anatomie europäischer Hölzer.* Verlag Paul Haupt, Bern and Stuttgart.

Teisseire, H., Fady, B. and Pichot, C. (1995) Allozyme variation in five French populations of Aleppo pine (*Pinus halepensis* Miller). *Forest Genetics* 2, 225–236.

Vallino, F.O. and Ventura, G. (1984) Dati archeobotanici dal Broglio: semi ed altri reperti [Archaeobotanical data from Broglio: seeds and other macroremains]. In: Peroni, R. (ed.) *Nuove ricerche sulla protostoria della Sibaritide* [*New researches on protohistory of Sibari land*]. Rome, pp. 272–280.

Zucconi, L. (1959) Del riconoscimento del legno dei pini mediterranei. Indagine microscopica [About the recognition of wood of Mediterranean pines. Microscopic investigation]. *Contributi scientifico-pratici per una migliore conoscenza ed utilizzazione del legno* [*Scientific-practical contributions for a better knowledge and utilization of wood*], CNR, Firenze 2, 1–10.

Xylology and Forest History

20

E. Corona

Facoltà di Agraria – DISAFRI, Università degli Studi della Tuscia (VT), Via San Camillo de Lellis s.n.c., 01100 Viterbo, Italy

Forest history can be reconstructed not only through written documents, but also by analysing the evolution and diffusion of the arts, professions and crafts related to forests and through the reference to toponyms, cults, creeds, superstitious beliefs and religions (Judaism, Christianity, Islam). Forests are characterized by the capability of recording abiotic and biotic environmental phenomena which have influences on the development and modification of their structures and components through time. Within such a framework, xylology applied to prehistoric and historic woodworks and to wood samples from plurisecular trees can greatly contribute to forest history studies. The anatomical, dendrochronological and densitometric characteristics of wood can give evidence about climatic events, forestry practices, aspects of wood product marketing and about the patterns of social structure reflected in the development of forest stands.

The interest in tree identification and classification has led researchers beyond the mere study of morphological and physiological features of the trees, towards the wider study of those aspects and information that directly or indirectly may help to highlight the history of forest ecosystems. For instance, the critical analysis of classical authors such as Theophrastus, Plinius and Dioscoris throws new relevant light on the forests of the past (Egger and Egger, 1988). Another example is the interpretation of Renaissance codes and miniatures which fosters the alliance between art and forest science.

With respect to the goals of forest history, it is effective to reconsider the forest cults that are widespread within ancient cultures. Almost all such cults subtend the protection of the trees. For instance, trees could neither be cut nor pruned according to Cato (*De agricultura*, 139). In the same way, it is interesting to consider the interpretation of the etymological meaning of the Latin words *nemus* and *lucus* in depth, and to examine the rites of old Semites, Assyrians, Egyptians and Persians, the statements of Deuteronomy (12:2), the meaning of the *Dendrophori* in the Roman Empire, the distinctions between happy and unhappy trees according to Gellius (X, 15, 15), the cult of trees by the Jews (see the hermeneutic of the Midrashes and Talmud) and finally the 22 Koranic Sures on trees (Di Nola, 1991). Evidence of the sacredness of the forest could still be found in the Middle Ages. Sacred forests lasted in Gallia and in Germany, imposing abnormal cults and superstitions that have sometimes also influenced the preferences for using certain woods. This happened in Italy too (e.g. Casentino).

Since ancient times, the relationship between the development of civilization and the environment has been complex. The history of this relationship is a manifold issue, rich in experiences and different ways of exploiting and controlling resources. In the economic history of the western world, the forest has been a main production centre of raw materials: e.g. timber for the construction of buildings, boats and harbours, and firewood for heating and cooking, for coal production and for metalworking. The study of wood uses is essential for the study of the history of man with his culture, as well as the climate.

The economic value of forests has led man to develop various management systems to balance the needs for agricultural development and timber and firewood with the need for nature conservation. Such economic value has been theorized since the times of Aristotle (*Politica* VI, VIII, 6). However, a notable contribution to forest history comes following the development of dendrochronological methods and techniques, with their specialization in densitometry, dendroclimatology, dendrotypology, etc. Data regarding tree growth, inter-tree competition, cuttings, climatic events, floods and so on, can be very informative to the historian, bringing effective elements of evaluation about events, composition and evolution of forest ecosystems and about harvesting, transport and marketing practices of the different ages.

For instance, xylo- and xylochronological research on Neolithic pile-dwelling settlements helped characterize the directions of their expansion, and showed the choice of alderwood as a raw material with a clear meaning from climatic and phytohistorical points of view. The 2300 poles of Cortalloid and Neuchâtel were proved to have been harvested in a clearcut even-aged forest; even the timberyard could be identified. Beyond the types of wood and the systems of harvesting and processing, the study of the wood from trees grown during the post-glacial climatic optimum has shown the preference of that time for using oak species, including holly oak, that were located in altitudinal belts where they can no longer be found.

The Egyptian pyramids still contain wood residuals of *Cedrus* and *Juniper* from Syrian and Lebanon forests and delicate wood elements made of the genus *Tamaryx*. Similarly, the preferences of the Greeks for firwood from the *Apollinis* and *Panachaica* forests have been highlighted by xylological studies. At that time, Aristotle (*Politica*, I, XI, 4) stressed the value and the impoverishment of the forests and the role of the 'Uloroi' (Fioravanti, 1991). The preferences of ancient Romans for silver firwood from the Apennines, often long-length logged through the narrow streets of Rome, have been documented too. In the first century BC, the larchwood from Germany and Gallia was used for harbour buildings in central Italy. Given its durability, *Cupressus sempervirens* wood was adopted for structural uses, and it was also distinctively used for sacred woodworks. This contributed to the diffusion of this species all around the Mediterranean basin. After the age when forests were safeguarded because of their wood production for arsenals, mines and housing, a vast deforestation began at the end of Roman Republican age, especially in central and southern Italy. This trend weakened during the Low Roman Empire when sheepfarming expanded, thus to a certain extent favouring natural forest regeneration, above all in the Po valley.

The study of woodworks can give reliable historical information from the ethnographic, philological, silvicultural and commercial points of view. The diagnosis of the wood species used for houses, harbours, tools, furniture, sculptures and paintings, together with their dendrochronological, densitometric and image analyses, can contribute to date such works; to determine preferences and processing techniques; and to reconstruct habits, cultural practices and commerces of the past.

In many cases, dendrochronological analyses of old built-up areas have solved the problems of dating buildings and answered questions about wood procurement, marketing and timber origin (Corona, 1997). A relevant case is the excavations in Novgorod where the dating of 3000 samples of Scots pine and Norway spruce allowed the reconstruction of the town development in time, and dated the sequence of fires in the town and the sequences of the wood paving of the streets (Eckstein, 1972). In Smolensk, 405 samples from 76 houses allowed buildings to be dated in a time span from the 12th to the 14th century. Similar dating methodologies were applied in Belosersk and Maganzeia, in many villages of England, Scotland, Poland, Sweden and Switzerland, and in the fortifications of Lusazia and Mecklemburgh (Eckstein *et al.*, 1983; Wrobel, 1992).

The type of timber chosen and its xylematic and xylochronological characteristics can offer interesting indications on habits and silvicultural practices, as well as on climatic and extraclimatic events that have influenced tree growth in the past, and often the history of mankind. For instance, through the study of the timber in the old village of Charavines Collettines

(France), a tree ring chronology of 155 years (886–1040 AD) was chronicled. The architectural types in the village varied regarding the availability of oak and beech woods: from the year 955 onwards for 50 years, tree growth was small for both species, and this fact has been related to glacier advancement and solar activity. Furthermore, in the timber of some buildings reductions of tree stem increments proved to be associated with changes of carbon isotopes (Bourquin-Mignot, 1994).

Many constructions in the Latium Appennines characterized by Medieval and Renaissance structures made of yew are of indirect interest from a silvicultural point of view. Today, the use of such a wood is unthinkable given that the yew tree is now a rare, protected species. In the hinterland of Messina many Medieval and Renaissance buildings are made of *Abies nebrodensis* Mattei wood. Today, this conifer is a strictly preserved tree species: only 37 trees remain, on Mt Scalone in the Madonie range. There is a similar situation regarding *Abies cilicica* Carr. in Syria and *Abies pinsapo* Boiss. in Spain (Sierra of Estopona), destroyed by intensive use for structural and inside woodworks in the past (Corona, 1994).

In Dublin, a study to elaborate an oak chronology spanning from the tenth to the 15th century showed that, up to the 14th century, old oak boles (age up to 420 years) were used. Towards the 15th century, the used boles showed fewer rings indicating that overall forest patrimony had become depleted and younger trees had to be cut. Relevant data on assortments and construction ages have been obtained from the wood of buildings and river facilities in Exebridge in southern England constructed during the 12th and 13th centuries. Useful data about cutting dates, wood assortment and seasoning have been obtained also from the London waterfronts along the Thames, constructed in the 12th and 13th centuries (Morgan and Schofield, 1978). Medieval works in beech and oak in Aachen, Mainz, Treviri and Mönchengladbach have been thoroughly examined by Hollstein (1980). Other studies on woodworks from northern Germany (e.g. the Cistercian cloister of Wienhausen) have been carried out by the Chair of Wood Biology at Hamburg University. Interesting information on management systems and timber origin emerge from a dendrochronological study on wood trusses in Lübeck: the result was that in the 15th to 16th centuries more sound systems replaced the raw *Plenterung* system developed in the 12th and 13th centuries. Imported wood from eastern countries used for specific works (e.g. the furniture and the organ of St Jakobi) could also be detected (Wrobel and Klein, 1996).

Xylological studies have also been carried out in Greece and in the Aegean islands, where the ancient builders used *Pinus nigra, Pinus leucodermis, Picea abies, Juniperus* species *plurimae*, *Abies* species *plurimae* and *Quercus* species *plurimae* (Kuniholm, 1983).

Xylological studies may also help to reconstruct lumber routes and markets. Investigations of oak woodworks of the 14th and following centuries from France and England showed that they were of Baltic origin. Curious evidence has even been obtained from the brackets of paintings by Rembrandt, Wouvermeer, Rubens and Molenhaer.

Following the comparison of tree ring chronologies, in the 8th and 9th centuries and especially after the impoverishment of Dutch forests, oak assortments were imported from Germany (more specifically from the Rhein area) to be used for architectural structures and harbour and river plants (Wazny, 1992). Quite a lot of woodworks in churches, palaces and other buildings of Florence and Rome were made of silver fir, whose different origins and processing techniques can be recognized by xylological methods. Similar considerations hold true for chestnut and walnut woodworks in Viterbo and Rome, which are characterized by unusual assortments (e.g. 'UF' beams, *ante litteram* 'head to head') (Corona and Lo Monaco, 1993). The larches from Cadore were used in all the arsenals on the Adriatic coast and even for the door coating of St Mark's in Venice in the 10th century (Pignatelli, 1996).

Studies carried out on 104 buildings in Jaffa, one of the most important Mediterranean harbours, resulted in the identification of more than 15 tree species, most from eastern Mediterranean regions (*Cedrus libani, Pinus nigra, Pinus brutia, Abies cilicica*) and from Europe (*Pinus silvestris, Abies alba, Larix decidua, Picea abies, Fagus silvatica, Quercus* sp., *Populus* sp.), some from non-European countries but procured through Europe (*Pseudotsuga menziesii, Cedrus atlantica, Tsuga canadensis*), and some grown locally (*Ceratonia siliqua, Quercus calliprinos*). The identification of the species, the types of assortments and, above all, tree ring analyses have helped to characterize the geographical origin of the timbers and even to identify the harbours of embarkation, e.g. Marseille for the cedars and oaks and Trieste for the spruce and larch. Together with other sources, these analyses helped to build a picture of the temporal sequences of emigration to the Middle East, and of the development of commercial activities, building style evolution, and the influence of different ethnic groups on the architectural typologies. Initially, Jaffa was expanded by the Ottomans and indigenous people, then by Jewish, Christian and Muslim immigrants. In some buildings the successive stages of construction can be clearly distinguished as a function of the timber used (Biger and Lipschitz, 1991).

An interesting dendrohistorical and dendrotypological study carried out in the French Giura on original and restructured buildings dating from the 17th century to today (Billamboz, 1990) evidenced the habit of using broadleaf timber (in particular oaks) up to the middle of the 19th century, when conifers (in particular silver fir and spruce) began to be preferred. Up to the 18th century, systematic differences between the certified dates of construction and the dendrochronological dates can be stressed in France: such differences can be explained because of the 'droit pour maisonner', which lasted up to the 18th century, according to which the same woodlot served for many constructions at different times. In France, the villages expanded quickly during the 17th century after the Thirty Years War that had depopulated the countryside and the mountains. In the 18th century, demographic growth, testified by the increasing number of houses, stopped temporarily but later continued. In the middle of the 19th century, the architectural types and styles and adopted wood species changed quite abruptly. In the buildings of the 17th century, the wood assortments, in particular the beams, were made mainly of very old oaks. Under Charles V (16th century) and Phillip II (17th century) the forests were growing well and the regeneration was flourishing, as is revealed by tree stem radial growth and by the synchronism of tree ring sequences. Later, forest policy changed, the Ordinance of Colbert (1669) was set and towards the end of the century the 'quart en réserve' management system was introduced. In the 18th century, under Louis XIV, the 'tire et aire' system was introduced, and in the French Compté the furetage system too – as can be seen from woodworks, the wood assortments originated predominantly from defective trees selected by severe criteria. In the meantime, population and herds increased, rangeland and hornbeam stands became widespread, oak forests become poorer, and the utilization of conifers increased. A wood supply crisis began in the 18th century according to Pluche in *Spectacle de la Nature*. In the entry Bois of his famous *Encyclopedia*, D'Alembert wrote that in France wood yield hardly satisfied the demand. In 1713, von Carlowitz published his famous *Sylvicoltura Oeconomica* (Corona, 1991).

In the Alps, block (*Blockbau*), bench-uprights (*Standbohlenbau*) and crossing (*Bundwerk*) architectural styles can be distinguished, marking different ages. The oldest technique is the *Blockbau*, characterized by simple carvings on the ends of larch, Norway spruce, silver fir, and Scots pine boles, squared and roughly worked by axe. Later, crossing architecture took place and then, in the 17th century, framed structures similar to the St Andrew cross were introduced. In that period, trusses were oversized in order to delay eventual destruction in the event of fire and to prevent the deterioration by vapours of stabling and/or from the smoke of open fireplaces. Roof architecture presented a parallel evolution: first, only roughly squared boles were used and then more finished frames, trusses and shingles (Sebesta, 1987; Zampedri, 1991).

Wood assortments used in the buildings of the Middle Ages can provide useful information on climatic history too. Traces of the early medieval warm period can be recognized in larches in the maritime Alps. In larch chronology by Brehme on buildings in the central Alps and Bavaria, increments twice as large as those during the 16th century can be recognized. This reduction was due to the well-known Little Ice Age climatic influence. Such climatic events characterizing the central and northern European environments are supported by well-defined signatures (Landshut, Doppeldepression, etc.) that often acted, in the same sense and with the same intensity, in the Mediterranean area too.

Sophisticated studies are carried out by means of densitometry, tomography and image analysis in order to characterize intra-annual climatic trends. In this respect, the second millennium has been studied mainly by Swiss, Swedish, German, English and also Italian research groups. These studies have focused on wood assortments from Norway spruce and silver fir; however, other species such as Scots pine, larch and oaks have been also investigated (Corona, 1992).

In the light of the above briefly described issues, it is clear that xylological studies and, more specifically, the dendrochronological ones, can provide an effective contribution to forest history (*sensu lato*). Through tree species identification, typological wood assortment interpretation and tree ring analyses it is possible to reconstruct not only the climatic events, but also the uses, marketing and transport practices of the past, and then to outline their eventual direct and indirect influences on contemporary forests.

References

Biger, G. and Liphschitz, N. (1991) Regional dendrohistory and timber analysis. *Mediterranean Historical Review* 6, 86–104.

Billamboz, A. (1990) Etude dendrochronologique et approche historique d'un village du Jura Comtois. *Dendrochronologia* 8, 99–117.

Bourquin-Mignot, C. (1994) Dendrochronologie du hêtre sur un site medieval des Prealpes Francaises. *Dendrochronologia* 12, 103–115.

Corona, E. (1991) Xilologia: implicazioni tecniche, storiche, filologiche. *Ann. Acc. It. Sc. For.* 60, 211–236.

Corona, E. (1992) La Dendrocronologia come strumento per lo studio delle variazioni climatiche. *Atti Conv. Lincei* 95, 113–128.

Corona, E. (1994) Il contributo della Tecnologia del legno nello studio della biodiversità. *Atti Occhi Verdi Sulle Foreste*, 63–65.

Corona, E. (1997) Il contributo della xilologia in alcune ricerche su manufatti medioevali. *Atti 1° Congr. Naz. Archeologia Medievale*, 9–11.

Corona, E. and Lo Monaco, A. (1993) Le strutture lignee di S. Maria in Forcassi. *Inf. Centro di Catalogazione Beni Culturali* 9, 7–10.

Di Nola, C. (1991) L'albero nelle religioni monoteiste. *De Arbore*, 429–534.

Eckstein, D. (1972) Tree ring research in Europe. *T. R. Bulletin* 32, 1–18.

Eckstein, D., Wrobel, S. and Aniol, R.W. (1983) Dendrochronology and archaeology in Europe. *Ord. für Holzbiologie*, 140, 1, 249.

Egger, H. and Egger, S.K. (1988) Dendrochronologischer Nachweis zur Herkunft von Bauholz. *Dendrochronologia* 6, 91–98.

Fioravanti, R. (1991) Sfruttamento economico e tutela nella società preindustriale. *De Arbore*, 199–207.

Hollstein, E. (1980) *Mitteleuropäische Eichenchronologie*. Ph. von Zabern, Mainz, pp. 1–273.

Kuniholm, P.I. (1983) Dating in the Aegean region. *Mitt. B. Forst und Holzw.* 141, 179–194.

Morgan, R. and Schofield, J. (1978) Tree ring and the archaeology of the Thames waterfront in the City of London. *BAR*, 223–238.

Pignatelli, O. (1996) La dendrocronologia in Italia: vari aspetti applicativi della disciplina. *Dendronatura* 17, 15–18.
Sebesta, G. (1987) Il legno nelle primitive abitazioni – case dell'uomo. *Linea Ecologica* 2, 50–57.
Stroppa, M. (1989) Analisi di strutture lignee provenienti da edifici storici della provincia di Verona. *Dendrochronologia* 7, 123–143.
Wazny, T. (1992) Historical timber trade and implications on dendrochronological dating. *Lundqua Report*, 331–333.
Wrobel, S. (1992) Dendrochonology with man used timber. *Lundqua Report*, 343–346.
Wrobel, S. and Klein, P. (1996) Dendrochronologie: ein Fenster in die Vergangenheit. *Forschungsreport* 1, 2–5.
Zampedri, G. (1991) Il legno nell'architettura del passato in Trentino. Atti Conv. Risorsa Legno, *Tecnologia, Architettura, Arte*. PAT, 101–112.

Bristlecone Pines and Tales of Change in the Great Basin

21

M.P. Cohen

Language and Literature Department, Southern Utah University, Cedar City, UT 84720, USA

In 1957, Edmund Schulman found individual bristlecone pines in the mountains of the Great Basin of Nevada and California that had lived for nearly 5000 years. Since then, the trees have been investigated primarily for the elaborate record their growth rings contain. Rings of Great Basin bristlecone pine (*Pinus longaeva* D.K. Bailey) have been 'read' closely, during a very brief era, with major consequences for theories of natural and human history. The idea of a tree as a text is itself an intricate ancient and modern social construction.

The history of the study of these trees reveals the problem of scale in reading natural history from natural systems. Pioneering dendrochronological studies – microscopic investigations of the interiors of selected trees – contrast with the studies of palaeoclimatologists, who have read global environmental changes out of tree ring data. Archaeologists altered their 'histories of civilization' as a result of this record. Beginning with Schulman, who was searching for old trees exhibiting 'longevity under adversity', investigators have transformed trees into figures pertinent to the human dilemmas of time and eternity, by literally inscribing human events on their rings.

Bristlecones are part of the 'first nature', and the records within the trees have provided a grounding of dating techniques like radiocarbon dating. Yet the use of bristlecone data has led to diverging paths in the methods of studying the intersection of forest history and human history.

 Methods and Approaches in Forest History (eds M. Agnoletti and S. Anderson)

1 Trees as texts

My recent book, *A Garden of Bristlecones: Tales of Change in the Great Basin*, is about one species of tree and what natural and cultural historians have learned from it. As one of my colleagues is fond of saying, environmental historians have only two tricks. If you say 'nature', they will try to show you that the story is about culture, and if you say 'culture', they will try to show you that the story is about nature. My stories are about nature, but also about how people know it, and consequently about culture.

As to the Great Basin and its forests: it is not a basin but an arid mountainous country, locked between the Sierra Nevada of California and the Rockies of Colorado, mostly in the state of Nevada. Many ranges rise to over 4000 m above sea level. The annual precipitation of Nevada is the lowest in the United States, about 23 cm $year^{-1}$. Winter snowpack on the highest peaks allows a set of floating forests on the mountains. The bristlecone pine is part of a timberline forest; it grows, usually with limber pine, at 2600–3500 m elevation in about 27 ranges of mountains. The tree itself is extraordinary.

What is remarkable about the Great Basin bristlecone pine (*Pinus longaeva* D.K. Bailey)? Not simply what scientists and cultural historians have gathered from it, but how recently this knowledge has been abstracted from the forests. In 1958, Edmund Schulman published an article not in a scientific journal, but in *National Geographic*, about the aged trees he found high in the White Mountains (Schulman, 1958). These trees were nearly 5000 years old. Age, in itself, does not constitute the value of the trees. It is remarkable, however, that humans left the old-growth bristlecones nearly untouched until the middle of the 20th century. A deferred knowledge of these trees became associated with the scientific tools of dating created in the emerging nuclear sciences.

Two things are remarkable in this regard. Earlier researchers looking for old trees sought monumental trees, like the *Sequoia* of the western slope of the Sierra, that live over 3000 years. The reason old bristlecone pines remained untouched for so long is that they seemed insignificant, and of no use. Secondly, the knowledge within the tree was acquired – as data – before the nature of the tree was fully understood. The Great Basin bristlecone was only identified as a species in 1970, after many of the tree ring studies upon it were completed.

In the context of knowledge, old trees are only valuable in so far as the record contained in them can be read. For instance, there are rings of creosote, a resinous desert shrub found in the Mojave Desert of California, that are over 10,000 years old, but their centres are gone, and so they do not contain a readable record. The point of investigating the tree rings in the bristlecone pine was that the record was readable. Reading begins with what is called 'chronology building', whereby the scientists, called dendrochronologists, establish a chronology counting back from the present, marked by the most recent growth ring of a living tree. Through a process of what is called 'cross-dating', the record of tree rings in pieces of old and dead trees can be used to extend the chronology beyond living trees. At present, the chronology contained in bristlecone pines comes close to the edge of the last ice age in the region, about 8600 years.

It is one thing to establish a chronology, and another to read the record from it. The record of these trees is in 'fine print'. Since these trees grow in such an arid region, there are approximately 120 growth rings in every radial inch of wood.

How humans read tree rings and also write on them, or inscribe human history upon them, is entirely a cultural matter. As an environmental historian who is also a literary critic, I have attempted to create – in my study of the study of trees – a hermeneutics of bristlecone reading. Chronology building can be taken as the first step, the creation of an uncorrupted text. The establishment of this text is entirely a result of processes of natural history; the reading and interpretation of this text is in its origin entirely a matter of the cultural history of science.

The scientists who first read the record of these trees worked at the University of Arizona, Tucson, in southern Arizona near the Mexican border. In the arid southwestern United States, the questions asked had to do with climate, and particularly with drought. In fact, the first famous use of dendrochronology was not with bristlecone pines, but with the wooden beams that were used by a prehistoric culture to create the abandoned cities spoken of as 'Anasazi' in the southwestern USA. By dating the logs, scientists were able to date these cultures, and give them a history, so to speak. The unanswered question about these cultures was, however, why they abandoned the region, and the 'assumed' answer to that question was taken to be climatic.

However, the first use of the bristlecone pine chronologies was in the 'calibration' of the radiocarbon dating technique for which Willard Libby received a Nobel Prize in 1960. Libby assumed constant levels of carbon-14 in the earth's atmosphere; he was wrong. For that and possibly other reasons, radiocarbon dating was insufficiently accurate to date and compare human artefacts, and it was only in the mid-1960s that the precise though primitive science of tree ring studies was used to correct the errors in Libby's subtle but inaccurate measuring system. As before, the result was turned toward cultural history. After 'calibration' – testing with real bristlecone wood samples – radiocarbon dating altered the cultural history of Europe by destroying the 'diffusion' model of cultural development in Western and Eastern Europe.

Dendrochronologists always sought not chronology but a record of climate. This record was largely assumed to be continuous and perhaps cyclical until a new ecology began to emerge in the mid-1980s, as a result of Pickett and White's (1985) *The Ecology of Disturbance and Patch Dynamics*, and what is now sometimes called 'chaos ecology'.

It was as a result of this, or in connection with this perspective, that tree ring scientists began to look at the discontinuities found in the growth rings of timberline trees in the White Mountains. In particular, there is something called a 'frost ring' that shows damage in the wood reflecting a disturbance in the timing of seasonal change. By a fortuitous chance, the scientist studying this phenomenon in timberline trees in the White Mountains acquired a colleague who was expert in the global chronology of major volcanic eruptions. The result was what is called 'correlation' of the two chronologies and the accurate indexing of major disruptive events, a list of over 30 of them, in the earth's climatic history.

Other data, some from ice cores, some from oaks dredged out of bogs in Ireland and Germany, and so on, have once again extrapolated this set of events to a set of global events. There have been many global events, as palaeoclimatologists call them – certainly not all caused by volcanic eruptions – that resulted in severe multiyear plunges in the biomass of plants growing on earth. The most obvious implication of this particular 'discovery', or more accurately 'correlation', is that tree ring records can be subjected to successive re-readings using different methods or modes of reading. As literary critics would say, tree rings constitute a text: a field for multiple readings. Local forests allow a reading of global climate history.

This is not a complete set of stories, but a fair sample of how they have been constructed. So briefly, what can be concluded from examining these stories?

2 Science

The study of bristlecone pines at the timberline in the White Mountains allowed the timberline to be drawn around the world. The timberline is a biotic boundary between environments in which trees can grow and where they cannot. As a result, it is an index of climatic conditions, and changes in this line indicate changes of global climatic conditions. In forest history, at least in this case, study of a microcosm leads to a macrocosmic history. But chronology is not about plot. Forest history is always about plot, and this line must have specific events inscribed upon it, as events with potential causes, that give the story structure and meaning.

Using not just the oldest trees, but the most stressed (those whose growth was limited by a short arid growing season), and taking cores from them, registering die-offs and recruitment of the forest, scientists still choose individual trees. The timberline, read as a line between the life and death of a species, is mapped by choosing individuals. The individual and even species are replaceable, but the movement of the timberline is real because it is created by the births and deaths of populations of trees and therefore accurately indicates actual changes in global climate. What the scientists were studying was an existential question, though they did not study it as such. The study of climate change, rooted in migrations of the bristlecone, is an existential question.

The movement of this line reveals changes in the conditions under which life flourishes on earth, and it is always studied in reference to human lives. The timeline is always rooted in the present, attempts to read the past with reference to the present, and seeks to project a possible future on the basis of the past.

Precipitation and temperature changes in the past have, for most of our history, been imagined to be gradual, and the recent tree ring record shows this to be largely true. There are, however, also signs of discontinuity. The frost ring and correlation with other data constitute a frightening existential problem of the discontinuity of life.

All study of the timberline attempts to recover past conditions relevant to our lives. This recovery has been both comforting and frightening. It is comforting because the trees have survived and have continued to grow for thousands of years in the same places, and have continued to evolve over millions of years in the same regions. It is frightening because the forests have survived under adverse circumstances and through major climatic disturbances that perhaps were disasterous to human populations in the past and certainly would be in the future.

3 Art

Science does not exist in a vacuum. It is informed by questions of the culture around it and informs that culture. Timberline forests also provide an aesthetic opportunity, and the implications of timberline forests on aesthetics can also be drawn around the world. The stressed trees represent something to humans of the gestures of life at the edge of its existence, between life and death. This is a real existential experience, and perhaps the bottom of all human experience.

Modern interest in the forms of destruction and the sums of destruction in paintings by cubists like Picasso and sculptures like those of Giacometti, has drawn artists and spectators to the timberline trees. The deferred interest in these trees is not just scientific, but aesthetic. Only when we were ready to ask certain questions were we able to see what was there. These trees were only present because we had found no previous use for them.

That this process has gone on in places largely unfrequented by humans means that humans now consider that they are recovering conditions of a past beyond their own, and rewriting or refiguring their own past through the use of these aesthetic forms created before people were conscious of them.

4 The shape and meaning of the perseverance of life

The history of life has often been described as a tree, and the Darwinian Tree has a distinctive shape. This as it happens corresponds, by pure coincidence, with the aesthetic shape of the Great Basin bristlecone pine. But the idea of evolution is also about limits and the changes these limits create in life. The bristlecone pine becomes, as a result, an objective correlative of human thought, both the occasion and the result of it, a part of the story we tell about ourselves and about our worlds, and the story that links our story to the world's stories.

In this sense, it is true, as Umberto Eco (1994) suggests in his *Six Walks in the Fictional Woods*, that woods are a metaphor for a narrative text because a forest is a garden of forking paths. These woodlands offer a particular narrative that only modern peoples were apparently able to appreciate, and so now as we wander in them we know that they have been waiting for us, and when we choose them we suspect that they have chosen us.

References

Cohen, M.P. (1998) *A Garden of Bristlecones: Tales of Change in the Great Basin*. University of Nevada Press, Reno.

Eco, U. (1994) *Six Walks in the Fictional Woods*. Charles Eliot Norton Lectures, 1993. Harvard University Press, Cambridge, Massachussetts.

Pickett, S.T.A. and White, P.S. (1985) *The Ecology of Natural Disturbance and Patch Dynamics*. Academic Press, New York.

Schulman, E. (1958) Bristlecone pine, oldest known living thing. *National Geographic Magazine* 113, 354–372.

Further reading

Alexander, T.G. (1987) *The Rise of Multiple-Use Management in the Intermountain West: a History of Region 4 of the Forest Service*. USDA Forest Service Report No. FS-399. USDA, Washington, DC.

Anon. (1958) Edmund Schulman, 1908–1958. *Tree-Ring Bulletin* 22, 2–6.

Anon. (1962) Andrew Ellicott Douglass, 1867–1992. *Tree-Ring Bulletin* 24, 2–10.

Arno, S.F. and Hammerly, R.P. (1984) *Timberline: Mountain and Arctic Forest Frontiers*. The Mountaineers, Seattle.

Axelrod, D.I. and Raven, P.H. (1985) Origins of the Cordilleran flora. *Journal of Biogeography* 12, 21–47.

Bailey, D.K. (1970) Phytogeography and taxonomy of *Pinus* subsection *Balfourianae*. *Annals of the Missouri Botanical Garden* 57, 210–249.

Baillie, M.G.L. (1995) *A Slice Through Time: Dendrochronology and Precision Dating*. B.T. Batsford, London.

Barbour, M.G. (1995) Ecological fragmentation in the fifties. In: Cronon, W. (ed.) *Uncommon Ground: Toward Reinventing Nature*. W.W. Norton & Co., New York, pp. 233–255.

Beasley, R.S. and Klemmedson, J.O. (1980) Ecological relationships of bristlecone pine. *American Midland Naturalist* 104, 242–252.

Betancourt, J.L., Van Devender, T.R. and Martin, P.S. (1990) *Packrat Middens: The Last 40,000 Years of Biotic Change.* University of Arizona Press, Tucson.

Botkin, D. (1990) *Discordant Harmonies: a New Ecology for the Twenty-first Century.* Oxford University Press, New York.

Bradley, R.S. (1985) *Quaternary Paleoclimatology: Methods of Paleoclimatic Reconstruction.* Unwin Hyman, Boston.

Brubaker, L.B. (1986) Forest disturbance and tree-ring analysis. In: Jacoby, G.B. and Hornbeck, J.W. (eds) *International Symposium on Ecological Aspects of Tree-Ring Analysis in Marymount College,* Tarrytown, New York. US Department of Energy, pp. 101–118.

Charlet, D.A. (1996) *Atlas of Nevada Conifers: a Phytogeographic Reference.* University of Nevada Press, Reno.

Connor, K.F. and Lanner, R.M. (1987) The architectural significance of interfoliar branches in *Pinus* subsection *Balfourianae. Canadian Journal of Forest Research* 17, 269–272.

Connor, K.F. and Lanner, R.M. (1991) Effects of tree age on pollen, seed, and seedling characteristics in Great Basin bristlecone pine. *Botanical Gazette* 152, 107–113.

Critchfield, W.B. (1977) Hybridization of foxtail and bristlecone pines. *Madroño* 24, 193–212.

Currey, D.R. (1965) An ancient bristlecone pine stand in eastern Nevada. *Ecology* 46, 564–566.

Douglass, A.E. (1928) *Climatic Cycles and Tree Growth.* Publication No. 289. Carnegie Institution, Washington, DC.

Eagleton, T. (1990) *The Ideology of the Aesthetic.* Basil Blackwell, Oxford.

Ferguson, C.W. (1968) Bristlecone pine: science and aesthetics. *Science* 159, 839–846.

Ferguson, C.W. and Graybill, D.A. (1983) Dendrochronology of bristlecone pine: a progress report. *Radiocarbon* 25, 287–288.

Fiero, B. (1986) *Geology of the Great Basin.* Max C. Fleishmann Series in Great Basin Natural History. University of Nevada Press, Reno.

Fritts, H.C. (1969) Bristlecone pine in the White Mountains of California: growth and ring-width characteristics. *Papers of the Laboratory of Tree-Ring Research,* Vol. 4. University of Arizona Press, Tucson.

Fritts, H.C. (1976) *Tree Rings and Climate.* Academic Press, London.

Fritts, H.C. (1991) *Reconstructing Large-Scale Climatic Patterns from Tree-Ring Data – a Diagnostic Analysis.* University of Arizona Press, Tucson.

Gates, D.M. (1971) An amateur botanist's great discovery: Dana K. Bailey and *Pinus longaeva. Missouri Botanical Garden Bulletin* May, 39–48.

Gleick, J. (1988) *Chaos: Making a New Science.* Viking Penguin, New York.

Gould, S.J. (1987) *Time's Arrow, Time's Cycle: Myth and Metaphor in the Discovery of Geological Time.* Harvard University Press, Cambridge, Massachussetts.

Graumlich, L.J. (1994) Long-term vegetation change in mountain environments. Palaeoecological insights into modern vegetation dynamics. In: Beniston, M. (ed.) *Mountain Environments in Changing Climates.* Routledge, London, pp. 167–179.

Graumlich, L.J. and Brubaker, L.B. (1995) Long-term records of growth and distribution of conifers: integration of paleoecology and physiology ecology. In: Smith, W.K.

and Hinckley, T.M. (eds) *Ecophysiology of Coniferous Forests.* Academic Press, San Diego.

Grayson, D.K. (1993) *The Desert's Past: a Natural Prehistory of the Great Basin.* Smithsonian Institution Press, Washington, DC.

Hall, Cl.A., Jr (1991) Natural history of the White-Inyo Range, eastern California. *California Natural History Guides*, Vol. 55. University of California Press, Berkeley.

Harmon, W. and Holman, C.H. (1996) *A Handbook to Literature*, 7th edn. Prentice Hall, Upper Saddle River, New Jersey.

Hiebert, R.D. and Hamrick, J.L. (1983) Patterns and levels of genetic variation in Great Basin bristlecone pine, *Pinus longaeva. Evolution* 37, 302–310.

Hirshboeck, K.K. (1980) A new worldwide chronology of volcanic eruptions (with a summary of historical ash-producing activity and some implications for climatic trends of the last one hundred years). *Palaeogeography, Palaeoclimatology, Palaeoecology* 29, 223–241.

LaMarche, V.C., Jr (1963) *Origin and Geological Significance of Buttress Roots of Bristlecone Pines, White Mountains, California.* United States Geological Survey Professional Paper No. 475-C. US Geological Survey, Washington, DC.

LaMarche, V.C., Jr (1968) *Rates of Slope Degradation as Determined from Botanical Evidence, White Mountains, California.* Geological Society Professional Paper No. 352-I. United States Government Printing Office, Washington, DC.

LaMarche, V.C., Jr (1973) Holocene climatic variations inferred from treeline fluctuations in the White Mountains, California. *Quaternary Research* 3/4, 632–660.

LaMarche, V.C., Jr (1974) Paleoclimatic inferences from long tree-ring records. *Science* 183, 1043–1048.

LaMarche, V.C., Jr and Harlan, T.P. (1973) Accuracy of tree ring dating of bristlecone pine for calibration of the radiocarbon time scale. *Journal of Geophysical Research* 78, 8849–8858.

LaMarche, V.C., Jr and Hirshboeck, K.K. (1984) Frost rings in trees as records of major volcanic eruptions. *Nature* 307, 121–126.

LaMarche, V.C., Jr and Mooney, H.A. (1967) Altithermal timberline advance in western United States. *Nature* 213, 980–982.

LaMarche, V.C., Jr and Mooney, H.A. (1972) Recent climatic change and development of the bristlecone Pine (*P. longaeva* Bailey), Krummholz Zone, Mt Washington, Nevada. *Arctic and Alpine Research* 4, 61–72.

LaMarche, V.C., Jr, Graybill, D.A., Fritts, H.C. and Rose, M.R. (1984) Increasing atmospheric carbon dioxide: tree ring evidence for growth enhancement in natural vegetation. *Science* 225, 1019–1021.

Lambert, D. and Muench, D. (1972) *Timberline Ancients.* Charles H. Belding, Portland, Oregon.

Lanner, R.M. (1983) *Trees of the Great Basin: a Natural History.* Max C. Fleishmann Series in Great Basin Natural History. University of Nevada Press, Reno.

Libby, W.F. (1955) *Radiocarbon Dating.* University of Chicago Press, Chicago.

Libby, W.F. (1963) Accuracy of radiocarbon dates. *Science* 140, 278–280.

MacArthur, R.H. and Wilson, E.O. (1967) *The Theory of Island Biogeography.* Princeton University Press, Princeton, New Jersey.

McPhee, J. (1981) *Basin and Range.* Farrar Strauss Giroux, New York.

Mastrogiuseppe, R.J. and Mastrogiuseppe, J.D. (1980) A study of *Pinus balfouriana* Grev. & Balf. *Systematic Botany* 5, 86–101.

Mehringer, P.J., Jr and Ferguson, C.W. (1969) Pluvial occurrence of bristlecone pine (*Pinus aristata*) in a Mojave Desert mountain range. *Journal of the Arizona Academy of Science* 5, 284–291.

Mirov, N.T. (1967) *The Genus Pinus*. Ronald Press Co., New York.

Mooney, H.A., St Andre, G. and Wright, R.D. (1962) Alpine and subalpine vegetation patterns in the White Mountains of California. *American Midland Naturalist* 68, 257–273.

Muir, J. (1918a) Nevada forests. In: *Steep Trails*. Houghton Mifflin, Boston, pp. 164–173.

Muir, J. (1918b) Nevada's timber belt. In: *Steep Trails*. Houghton Mifflin, Boston, pp. 174–183.

Myers, A.A. and Giller, P.S. (1988) *Analytical Biogeography: an Integrated Approach to the Study of Animal and Plant Distributions*. Chapman and Hall, London.

Nydal, R. (1966) Variations in C14 concentration in the atmosphere during the last several years. *Tellus* 18, 271–279.

Peattie, D.C. (1991) *A Natural History of Western Trees*. Houghton Mifflin Co., Boston.

Renfrew, C. (1973) *Before Civilization: the Radiocarbon Revolution and Prehistoric Europe*. Alfred A. Knopf, New York.

Schama, S. (1995) The verdant cross. In: *Landscape and Memory*. Alfred A. Knopf, New York, pp. 214–226.

Schulman, E. (1954a) Tree-rings and history in the western United States. *Economic Botany* 8, 234–250.

Schulman, E. (1954b) Longevity under adversity in conifers. *Science* 119, 396–399.

Stokes, M.A. and Smiley, T.L. (1968) *An Introduction to Tree-Ring Dating*. University of Chicago Press, Chicago.

Suess, H.E. (1970) Bristlecone pine calibration of the radiocarbon time scale. In: Olsson, I.U. (ed.) *Radiocarbon Variations and Absolute Chronology*. John Wiley & Sons, New York, pp. 303–311.

Trimble, S. (1989) *The Sagebrush Ocean: a Natural History of the Great Basin*. Max C. Fleishmann Series in Great Basin Natural History. University of Nevada, Reno.

Vasek, F.C. (1980) Creosote bush: long lived clones in the Mojave Desert. *American Journal of Botany* 67, 246–255.

Webb, G.E. (1983) *Tree Rings and Telescopes: the Scientific Career of A.E. Douglass*. University of Arizona Press, Tucson.

White, R. (1991) *'It's Your Misfortune and None of My Own': A History of the American West*. University of Oklahoma Press, Norman.

Wright, R.D. and Mooney, H.A. (1965) Substrate-oriented distribution of bristlecone pine in the White Mountains of California. *American Midland Naturalist* 73, 257–284.

Zavarin, E., Snajberk, K., Bailey, D.K. and Rockwell, E.C. (1982) Variability in essential oils and needle resin canals of *Pinus longaeva* from eastern California and western Nevada in relation to other members of subsection *Balfourianae*. *Biochemical Systematics and Ecology* 10, 11–22.

22

Biodiversity and Forest Management: From Biodiversity to Geochronodiversity

P. Arnould, M. Hotyat and L. Simon*

Center for Biogeography-Ecology, UMR 8505, CNRS/ENS, Fontenay-Saint-Cloud, France

Biodiversity, development or long-lasting management and ecocertification form emerging and related topics in the field of forest research. They are all recent; they first appeared as concepts in the scientific field and were then adopted by politicians after the Conference of Rio in 1992. The term of biodiversity appears to be very complex for many reasons:

- Different definitions of the concept. There are many biodiversities (genetic, specific, ecosystemic, landscape, etc.) and researchers do not agree which level takes priority. Is it possible to study one aspect of biodiversity while neglecting others? The question is complicated by natural diversity coupled with cultural diversity.
- Problems of space. Environments with high levels of biodiversity are called natural (grasses, wastelands, meadows, forests, etc.). What is the contribution of biodiversity to the development of the forest: from bare soil to the old self-regenerating forest? What should we think of monospecific forests (oak or beech forests, poplar or coniferous plantations)? Is monospecific forest a synonym of impoverishment of biodiversity?
- Time questions. In France, as all over Europe, forests are not only ecosystems but also social products. The forest has a long geological and biological history, covering the evolution of life, continental drift

* With the collaboration of: C. Benfergani, E. Boisvert, A. Da Lage, V. Godard, and W. Winckler. Introduction and Section 1: P. Arnould; Section 2: L. Simon; Section 3 and conclusion: M. Hotyat.

 Methods and Approaches in Forest History (eds M. Agnoletti and S. Anderson)

and a short human history, from just after the Neolithic where man uses space and its products. How can we understand the different traces of ancient civilizations?

- Management questions. All forests have a past where management choices – conscious or spontaneous – ways and customs, perceptions and representations, the legal framework and economic and social pressures often played a major role. The end of coppices and full-grown forest trees, the demand for timber, firewood, wooden parts for boats and lumber are all important elements which help us understand the present biodiversity. The replacement of wood by coal, and after that by mineral oil, rural depopulation, the industrial revolution, the sale of national goods, and different management styles are also historical parameters of importance.
- Questions of appropriation. The historical appropriation (royal, seigneurial, clerical, communal forests, etc.) as well as the present appropriation of public, private and communal forests have to be taken into consideration to understand better the various problems of biodiversity.

To study simultaneously 'biodiversity' and 'forest management' you can not only study natural information about fauna and/or flora and technical choices. Our research aims to be deliberately wide open and interactive, which presupposes clearly defined concepts, methodology, techniques and tools. This methodology, created from private or public experimental forest sites spread over the Région Parisienne, is then tested in various experimental sites in Poland and Germany.

1 Introduction

Biodiversity is everywhere around us, directly perceived in plants and animals, but also present with ever increasing frequency in debates, in newspapers, in reviews, in conferences and in television and radio programmes. Biodiversity has become one of the key words in the environmental discourse of the last decade of the 20th century and into the 21st century. Along with sustainable development it has become a fashionable term, ubiquitous, unavoidable, in some way 'scientifically correct'. The world of forestry has seized hold of the term and now uses it at a remarkable rate of consumption.What, however, does this term really cover? Is it a new concept or an ancient idea? Is biodiversity the same for naturalists as it is for economists? Do agro-alimentary firms have a different vision of biodiversity from that held by rural groups? Can the question of biodiversity on a local scale be approached in the same way as it would be for a global perspective? How can one go from static inventories to dynamic studies comparing fixed situations at given moments, then to a truly diachronic approach through the fusion of geographic data and the integration of the dimension of the passage of time?

To approach biodiversity in biogeography one needs to constantly have three concerns in mind:

1. Firstly, that of taking account of the 'bio': the quality of living creatures, their groupings, interrelations and dynamics.
2. Then the need to situate living creatures in all the dimensions and on all the scales of the 'geo': the distribution of living creatures on the surface of the globe, from the ecological niche to the biome, addressing the question of places, territories, fluxes, networks, landscapes and all forms of spatial organization.
3. A commitment to respect the etymology of 'graphy': all that is linked to pictorial or written representation, or more generally to all forms of the representation, perception and expression of current or past biological and geographical knowledge. Texts and images constitute two major sources of this vast cultural domain. Epistemology and the history of sciences both play significant roles.

This biogeography is both natural and cultural, respectful of the parameters of time and space in the aim of enriching the concept of biodiversity, often too narrowly naturalist, by including it in the larger concept of geochronodiversity.

2 Biodiversity: an ancient concept made contemporary

From a schematic perspective, it is possible to pinpoint two key moments in recent developments as crucial to the emergence of the concept of biodiversity. The first was the publication of E.O. Wilson's *BioDiversity* in 1988, and the second, the Rio Conference in 1992 and the elaboration of the Convention on Biodiversity. Since then, scientific writing on this question, and its associated by-product of media popularization, have witnessed an extraordinary proliferation. Foresters quickly found themselves on the front line dealing with this new global problem of the environment. How has it come about that biodiversity should gain such a primacy? Is this really a genuinely new question? What are its qualitative achievements and what value should be accorded to it? Is it possible to form a periodization of the questions of biodiversity?

2.1 Millennia of studies on biodiversity

Biodiversity is not really a new idea. Evaluating the number of living beings on our planet is a task that appears to have already been accomplished; the systematic researcher in this field is an endangered species. The first to carry out a survey of living beings in the Judaeo-Christian tradition was Noah. He gathered on his ark all the animal and vegetable species threatened by the flood. Alas, the Bible does not provide a precise list, but simply indicates a systematic embarkation of a male plus female couple of each species (Genesis 6, 19, 20). In the Middle Ages and the Renaissance the paintings of terrestrial paradise display the diversity of the living. The most famous and the most frequently reproduced, are those of Jan Brueghel de Velours. They overrepresent mammals and birds; the size and beauty of the plants and animals has priority over strictly quantitative preoccupations.

Two major works of Greco-Roman civilization were to serve as references on the subject of biodiversity for centuries to come. The first of these was by the Greek, Theophrastus, one of the founders of botanical science. Born in 372–371 BC in Eresos, a small most westerly town on the isle of Lesbos, Theophrastus followed Plato's teaching in Athens before becoming the understudy of Aristotle who was 12 years his senior. Moreover, after the exile and death of Alexander's pedagogue, he succeeded Aristotle as the head of the Lyceum. He taught several thousand followers and published approaching 220 works on subjects ranging from logic to rhetoric or politics. His major work on botany dates from 314–313 BC. It is sometimes called *A History of Plants*, sometimes *Researches on Plants* (Theophraste, 1988), short and practical titles, proposed *a posteriori* as a name for the course taught at the Lyceum, refined and improved over several years and which, like many Greek works, did not in fact have a precise title. This is one of the founding works of European botany. In it, vegetation is observed described and classified into grasses, trees, bushes and shrubs, and also according to other parameters, such as aromatic, alimentary, medicinal or harmful plants. It is clear, even from these initial scientific works, that biodiversity is not a neutral notion. It is considered and organized in accordance with its potential uses for man.

The other major contribution from Greco-Roman civilization is the work of Pliny the Elder. Born at Como in AD 23, he died at Naples in AD 79, during an eruption of Vesuvius. Pliny the Younger, his adopted son, was to be his principal biographer. His *Natural History*, in 37 volumes and published after his death, is a giant compilation of all the knowledge of the time on natural matters (Pliny the Elder, 1950). Of the 37 volumes, six are devoted to the study of plants. A large proportion of the names used in contemporary terminology were established at this time. Mythology plays a major role in this process: Achilles, Adonis, Daphne, Narcissus, Nymph and Hyacinthus, for example. The books devoted to animals start with the biggest and strangest: elephants, lions, panthers and hippopotamuses, and then proceed to an inventory of birds and finish with insects (. . . spiders, crickets, grasshoppers, ants . . .), asking such surprising questions as: Do they breathe? Have they blood?

The work of Pliny the Elder, following that of Theophrastus, contributed to the establishment of the technique used for biological survey as being that of an inventory of an encyclopaedic nature. However, both also contain questions and curiosities of a historical, geographical and social order. But, none the less, the infernal mechanism of biological enumeration had been set in motion. While experiencing a deceleration, this technique continued during the Middle Ages, due especially to the contribution of Arab thinkers such as Avicenna and El Biruri. The invention of the printing press and the discovery of America gave a fresh impetus to surveys of living beings. Gaston Bonnier, from as early as 1920, in *Le Monde Végétal*, and then more recently Hervé Le Guyader in his 1998 thesis on 'Theories and history in biology', both provide precious information on the constitution of inventories and on the progress of classifications and biological knowledge which accompany these works of the enumeration of the living.

2.2 A debatable total figure

The current cumulation of these millennia of calculating the number of living beings, sagas of extraordinary personalities, of moments of rapid enrichment of knowledge and periods of stagnation, has resulted merely in totals, that are partial or synthetic, and have been more or less rapidly cobbled together. In the first chapter of the third tome of his monumental *Treaty on Physical Geography*, entitled *Biogeography*, Emmanuel de Martonne (1955) states with some assurance the number of living beings surveyed up to the beginning of the 20th century. The same task is taken up by Wilson in 1988 who arrives at the figure of 1,400,000 identified living species, of which 300,000 are plants. This figure seems to be relatively well founded. From the 100 species of plants described in the work of Yen-Ti in China in 2200 BC, through the 700 types of vegetation recorded by Dioscoridus, a successor of Theophrastus in AD 64, and the 8000 species of plants and animals codified by Carl von Linné in 1758 in his *Systemae Naturae*, the work of the systematic researchers has enabled an enormous leap forward in the knowledge of biodiversity. However, the accounts of variable reliability, founded on extrapolations backed up with arguments of variable thoroughness, have led to the potential total figure of biodiversity being estimated within the enormous range of 3 to 100 million. Biodiversity specialists, in their legitimate desire to justify the pursuit of their works on inventories, treat such numerical estimations with scepticism, claiming them to be ill founded. How can scientists who propose such a wide and approximative range then expect to be convincing when they come to talk about the disappearance of species? The exaggeration in the estimates of species still to be discovered, discredits the evaluations of species that are either extinct or in the process of becoming extinct. Is it certain that the reference is always to real biodiversity and not virtual biodiversity? For instance, to proclaim the disappearance of 100 species a day in a UNESCO (Connexion, 1992) publication is scarcely credible!

2.3 One or some biodiversities: propositions for historical markers

Whatever the outcome of the intense current debates on the preservation of biodiversity, it seems to us possible to propose a historical periodization of research on the diversity of living beings. Everyone seems to agree that during more than two millennia, up to the 19th century, the essential task of generations of botanists and zoologists has been the study of specific biodiversity. The end of the 19th and beginning of the 20th century witnessed the establishment and development of studies on stational biodiversity. Phytosociology, pedology and ecology permit the characterization of units of biodiversity at a higher level. During the middle of the 20th century, landscape biodiversity, a still more general level, was explored. This is a major aesthetic and social development applied to an object at the intersection of the mineral, the living and the constructed, a source of differences of perceptions and multiple rivalries. In contrast to this global conception of landscape biodiversity, there were significant developments at the end of the 20th century in the notion of infraspecific biodiversity, of a molecular or genetic nature. This concept is all the rage in the research programmes of all laboratories that consider themselves up to date in large industrialized countries. It is also the object of ethical debates surrounding transgenetic individuals and genetically modified organisms, and in addition, of economic questions of considerable importance for the control of the rights to animals and plants capable of providing enormous commercial profits.

In this light, biodiversity appears as a multiform and fluctuating concept. Trapped for a long time within the domain of specific inventory, it now extends well beyond that, both in the direction of the infinitely big and the infinitely small. Biodiversity can no longer be thought of in terms that are uniquely natural. Its field of evaluation has to include the integration of social, legal, economic and ethical factors. Its concrete study demands new procedures where the notions of time and space play a decisive role.

3 Three large European biomes in the light of the problem of biodiversity

Thinking about biodiversity in its multiple spatial and temporal dimensions demands the taking into account of the 'ecogeographical' context of each of the units studied. The three examples below provide an illustration of this procedure.

3.1 On the interface of specific and stational biodiversities: an old forest area in the plains of France

The forest of Saint-Gobain, situated to the north-east of the Paris Basin, provides a pertinent example of an old forest area established in the region for several centuries. The structure of its vegetation is marked by the importance of forest architecture whose slow evolution is principally controlled by the operations of forest management. The remarkable biodiversity of this forest mass, characterized by the presence of numerous species of chronological or ecological interest, is the result of two essential factors.

1. The position of the forest is on the frontier between two large phytogeographical units, the Atlantic and the Middle European domain. Numerous species belonging to one or other domain are present here and participate in the large specific biodiversity of the area. The rare species are joined by species in a situation of chorological marginality, whose interest is increased by the fact that their position here is an exceptional one.
2. The biodiversity is also a reflection of the stational mosaic that characterizes the area. The topographic dissection and the diversity of geological and soil types explain the large variety of biotopes. Beside the central climax, of oak and beech, numerous stational climaxes can be distinguished, most often characterized by original flora, as is proved by the abundance of submountainous species on the northern slope or even more so by the presence of certain boreo-arctic species amongst the peatbogs of the valley bed.

Ecological and phytogeographic conditions play the essential roles in conferring such richness on this old forestry area. The permanence of the area under forest conditions explains the relative monotony of the vegetation structures, in which can be found only the nuances of different forestry types. Biodiversity here is only linked to the variety of structures in a secondary way. It is only possible to observe the presence of certain rare species in relation to certain sylvogenetic phases (felling, for example) or certain types of treatment (the opening of the arborescent level in certain specialized cuts). In an essentially forestry milieu, by a definition a milieu of minimal activity, biodiversity remains above all a domain of specific richness, linked to a situation and to ecological conditions that are exceptional.

3.2 The dynamics of vegetation structures on the origin of landscape biodiversity: the mountain of Lure in the southern Alps

The case of the mountain of Lure, situated in the southern Alps, requires a different approach to biodiversity. Here, the forms of vegetation result from a recent evolution marked for more than a century by progressive forest dynamics that are spontaneous or voluntary. The more or less recent character of the evolution and the more or less rapid nature of these dynamics explain the juxtaposition on this mountain of Mediterranean affinities, of open milieus and forest areas in the process of closure. Thus, several types of biodiversity can be apprehended:

1. In the most anciently established forest areas are to be found, as was the case in the Saint-Gobain example, a biodiversity linked to notable stational or phytogeographic conditions. This is the case for the area of beeches on the northern slope of Lure, noted by Laroussinie (1988) for its abundance of rare species.
2. The biodiversities resulting from the mosaic of vegetation structures and environmental dynamics are older. Certain noteworthy species, such as the Scotch elm (*Ulmus glabra* Huds.), are localized in the ravines of the northern slope where morphological activity has slowed the establishment of a more banal climactic formation. Many noteworthy species equally are restricted to the semi-open forest formations, such as the summit beech groves rich in fritillary (*Fritillaria involucrata* All.). Even more obvious is the presence of an area of *Genista radiata* L. situated outside its usual area of habitation. This formation is the result both of the Quaternary era heritage and the interruption of the return of forestry caused by the use of the mountain as pastureland.
3. The diversity of vegetal formations is not only the cause of the biodiversity, it also becomes an element within it. The elements of the landscape are thus considered in themselves as constitutive of biodiversity. The beech area of the southern slope represents, as a landscape unit, a remarkable aspect of biodiversity on a regional scale where this type of formation is normally confined to northern slopes. The landscape mosaic with alternating areas of moors, grass areas and high formations, constitutes, in the same way as the presence of remarkable species, an important element in biodiversity.

Environments marked by strong recent changes demand an approach that goes beyond the traditional one of biodiversity viewed simply as specific richness. They require that the landscape and historical dimensions of an environment be taken into consideration, and that scales of biodiversity be combined, from species or subspecies to the mosaic of formations.

3.3 Tools for the approach to biodiversities: the forest of Bialowieza (Poland) and its surrounds

The contribution of satellite imagery and geographic information systems (GIS) to the analysis of biodiversity can be understood through a study carried out on the forest area of Bialowieza (the 'World Reserve of the Biosphere', MAB programme, UNESCO) and the young private forests which surround it. Although the extraordinary richness of the flora and fauna which characterize the area have already been the object of a multitude of studies, new information can be contributed with the aid of satellite imagery and GIS.

A first research line has made possible the presentation, from SPOT images, of the diversity of the forest structures of the area. In combination with ecological data integrated in the GIS, the analysis of the structural diversity of the forests enables the localization of the zones that are potentially the richest, those where diversified forest architectures combine with remarkable ecological conditions.

The analysis of landscape dynamics based on images taken at an interval of some years has also permitted the pinpointing and characterizing of thresholds and forest ecotones, whose ecological interest is manifest. Field studies were also indispensable in the checking of the typologies of ecotones defined from satellite imagery (Geroyannis and Simon, 1997). Finally, the statistical treatment of images taken from the classification makes possible the characterization of the structures of the landscape as a whole, the definition of indices of homogeneity/heterogeneity, and the analysis of combinations present in the landscape (size and form of landscape units and arrangement of different units amongst themselves). To consider biodiversity in its geographical dimension thus opens new research perspectives which make use of satellite imagery and GIS.

4 Towards a global approach to biodiversity: a methodological proposition

As has been demonstrated, biodiversity can be approached in different ways. Each has its methodologies and provides information at different levels: species, silvosystems and landscapes. These can all lead to cartographic documents: the localization of species, maps of forest typologies, or maps of landscape structures and dominant species. These procedures make use of comparative statistics in carrying out independent treatments for each date 'with the aim of the recognition of objects through comparable terminology' (Geroyannis and Simon, 1997). Landscape structures and the changes in landscapes can thus be evaluated through the intersection of individually classified images.

4.1 Towards procedures of fusion

These procedures can be further improved by increasing the number of dates, by overlapping scales and by multiplying map comparisons. But the information will still be segmented. The objective of these fusion procedures is to try to reunite the entirety of the facts in order to combine the three levels – specific, stational and landscape – to integrate several dates in order to analyse changes, and to obtain, finally, a typology of dynamics. Such a procedure may use several fusion procedures such as vectorial fusion (Fig. 22.1a), decisions (Fig. 22.1b) or geometrics. The first procedure reworks the images in a common geography, and then extracts for each pixel 'a value through the combination of its values in different frequency bands, in order to obtain a fusioned signature

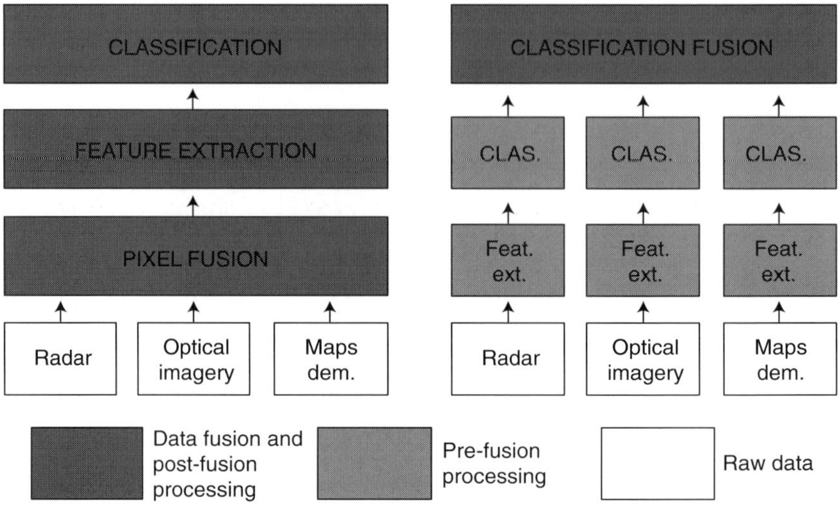

Fig. 22.1. Main data fusion techniques: (a) pixel-based fusion methodology, and (b) classification-based data fusion (from Liège, 1997).

of pixels which engenders a new hybrid radiometric value' (Liège, 1997). The second effects the fusion once the analyses have been carried out for each product. In this case, each image is worked on independently and only the classifications are fusioned. This procedure can be useful in the analysis of spatial changes within the framework of a multitemporal analysis. But, although the procedure is simpler, the interpretation of the information provided by each source is not always an easy task. Finally, geometric fusion, that most frequently performed, is not strictly speaking a fusion but a reworking of different cartographic documents and images. The latter is more pertinent, if the researcher is more interested in the morphological modification of objects than in their radiometric content.

These different procedures seem extremely seductive since they permit complementary information from different sources to be brought together. But, in order to really understand specific biodiversity, it is none the less necessary to make use of cartographic documents elaborated from rigorous fieldwork. In order to evaluate the transformation of landscapes over a longer period, it is indispensable to integrate historical documents with frequent problems of geometric fit. Similarly, in order to comprehend landscape diversity, aerial and land photographs are essential. Although it is easy to draw up a list of sources and documents used at each level of study, one fundamental problem remains: how to associate all these eclectic documents? Especially since, at the outset, the fusion of data brought together satellite data and aerial photographs. But, the objective of the research to associate biodiversities with each other and thus the specific sources of information to which they relate, demands that diverse sources be associated and that a still more complex methodological procedure be constructed.

4.2 From one fusion to another ...

This approach (Fig. 22.2) can be subdivided into three stages: a first phase of collecting information which includes field data, optic imagery including data from satellite and aerial photographs, radar, the whole gamut of cartographical documents and a numerical model of the terrain. The data are then submitted to different treatments allowing geographical corrections to be carried out and the establishment of a georeference system. At the same time, the field information will be structured in order to create a meaningful factual base. Then, the images will undergo treatments such as automatic or semi-directed classifications and the calculation of indices of vegetation, using different filters. Once this phase has been finished, fusion can begin, either by combining data from two sources at a time, or by fusioning several sources simultaneously. Whatever the type of fusion chosen a series of thematic maps are elaborated which make it possible to enter a new GIS database at a later date. These different layers of information can also be put on a digital elevation model (DEM) which can then serve as a base for a simulation.

The full use of this methodological chain presents biogeography with the advantage of establishing a coherent system based on information coming from different sources, of passing from a dynamic of static comparison to an integrated approach giving an indication of spatio-temporal modifications (Hotyat *et al.*, 1999). This can also make possible the construction of simulations with a view to dealing with the three biodiversities in a synchronic way. Perhaps this might be a way of genuinely apprehending 'geochronodiversity'!

Even if this methodological description appears simple and highly seductive, it possesses a certain number of constraints such as: the considerable mass of information demands very powerful equipment, time-consuming procedures and a significant cost. In order to evaluate its efficiency, it would be necessary to test it in other similar spaces; and in order to make it cost effective, the automatization of certain steps of the procedure is needed, which will require another significant phase of research development.

In conclusion, even if this procedure does not have as its goal the definition of specific biodiversity, although it does integrate this, it is none the less of genuine interest for the study of structural variations, the distribution of dominant trees and spatiotemporal modifications of the landscape on different scales. In addition, it offers several advantages including those of being reproducible, even of being automated; of being transposable for the study of similar environments; and permitting a genuine tracking of stational and landscape dynamics. This procedure has proved to be efficient in a restricted area of several hundred hectares, but now needs testing on larger areas.

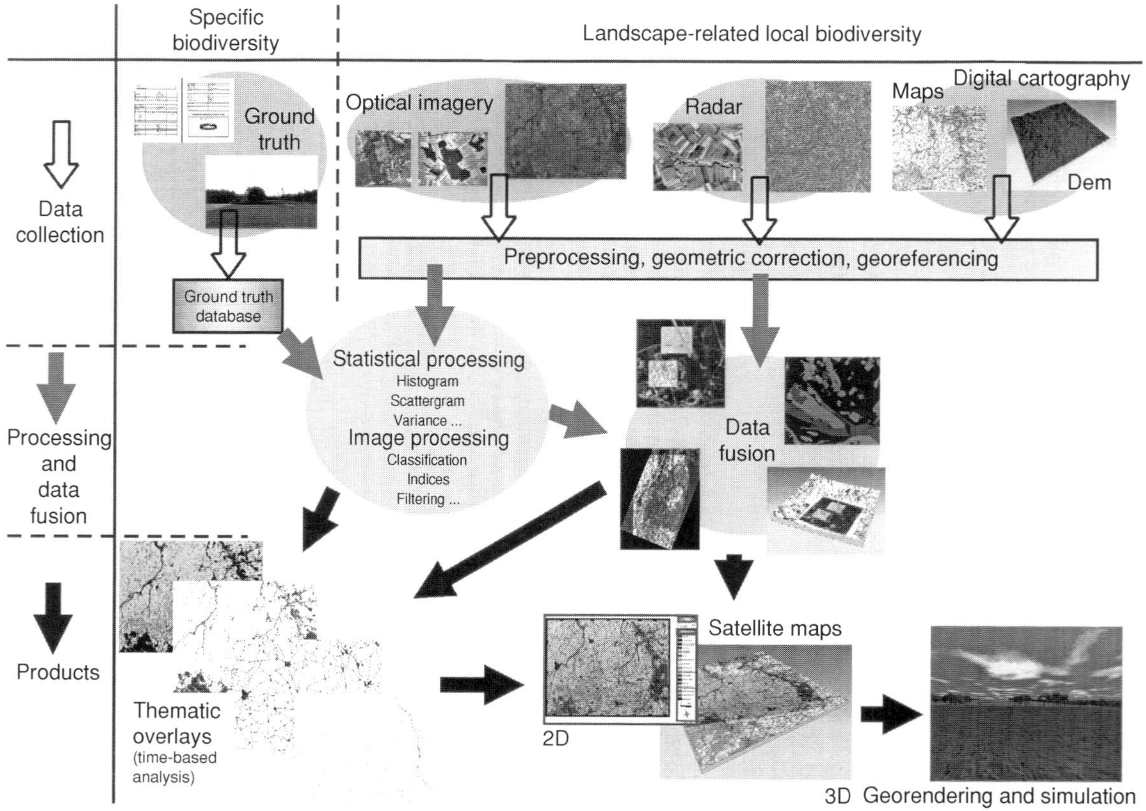

Fig. 22.2. Process methodology showing the three main stages (from Liège, 1997).

References

Arnould, P. and Blandin, P. (1996) *Devenir des îlots boisés dans les plaines de grande culture. L'exemple du Gâtinais nord-occidental.* Rapport final du programme thématique Systèmes écologiques et actions de l'homme. CNRS, Paris.

Aubertin, C. and Vivien, F.D. (1998) *Les enjeux de la biodiversité.* Economica, Poche Environnement, Paris.

Bonnier, J. (1920) *Le Monde Végétal.* Ernest Flammarion, Paris.

Bournérias, M. (1995) Le concept de rareté des espèces végétales (France et Europe occidentale). *Cahiers des Naturalistes* 51(2), 49–57.

Chauvet, M. and Olivier, L. (1993) *La biodiversité, enjeu planétaire.* Editions du Sang de la Terre, Paris.

Connexion (1992) *La Biodiversité: un Thème d'une Importance Croissante dans l' Education Environmentale.* UNESCO, p. 1.2.

Falinski, J.B. and Mortier, F. (1996) Biodiversité et gestion durable des forêts en Europe. In: *Revue Forestière Française.* Numéro spécial 'la gestion durable des forêts tempérées', sous la direction de Christian Barthod. 86–116.

Geroyannis, H. and Simon, L. (1997) *Télédétection et reconnaissance de structures paysagères, mise au point méthodologique.* Photo Interprétation, Editions ESKA, 35–42, 57–59.

Hotyat, M., Galachet, M. and Liege, F. (1999) Structure et dynamique des îlots boisés: intérêt de l'étude multicapteur et de l'analyse diachronique. L'exemple du Gâtinais occidental (France). *Paysages agraires et environnment, sous la direction de Stanislas Wicherek.* CNRS Editions, pp. 357–370.

Laroussinie, O. (1988) *Les patrimoines naturels forestiers des Alpes de Haute-Provence.* ENGREF, Nancy.

Larrère, R. (1998) Biodiversités. *Forêt Méditerranéenne* 14, 3–10 [extract from Larrere, C. and Larrère, R. (1997) *La crise environnementale.* INRA Editions, pp. 145–149].

Le Guyader, H. (1998) *Théories et histoire en biologie.* VRIN, Paris.

Lévêque, C. (1997) *La biodiversité.* Presses Universitaires de France, Paris.

Liège, F. (1997) *Gestion de l'espace par analyse multisources de l'information géographique.* PhD thesis, Université Paris-Sorbonne.

Linné, C. von (1758) *Systemae Naturae.* Leyden.

Martonne, E. de (1955) *Biogéographie* [*Biogeography*]. *Traité de Géographie Physique* [*Treaty on Physical Geography*], Vol. 3, 7th edn. Paris, pp. 1061–1541.

Pliny the Elder (1950) *Histoire naturelle.* Budé, les 'Belles lettres', Collection des Universités de France, 37 vols, Paris.

Simon, L. (1988) *Recherches biogéographiques en forêt de Coucy-Basse (Aisne), cartographie thématique à grande échelle d'un massif forestier.* Thesis, Université de Paris.

Théophraste (1988) *Recherches sur les plantes.* Budé, les 'Belles Lettres', 2 vols, Paris.

Wilson, E.O. (1988) *BioDiversity.* National Academy Press, Washington. [French edition (1992) *La diversité de la vie.* Odile Jacob, Paris.]

Temporal Differences in Forest History at Two Sites in Eastern North America

23

G.B. Blank

Department of Forestry, North Carolina State University, Raleigh, NC 27695, USA

Two forested sites provide a contrast in studies of forest history. The sites are located 460 km apart along the American east coast, with one on the eastern edge of the Carolina Piedmont and the other in the central Appalachian mountains. Settlement at both sites began in about the same time period, but proximity to transportation, industrial development, and therefore markets, differed considerably. The timing and types of disturbance experienced near these sites differed as well. Factors affecting research into the history of forest impacts at these sites are their proximity to the researcher and funding to support studies. One study has remained general in scope, whereas the other study has led to detailed site analyses and specific results generating a plan for adaptive management and ecosystem research.

1 Introduction

Differing rates of early settlement and development along North America's east coast led to varied impacts on forested landscapes. Overall, of course, disturbances occurred nearly everywhere. Yet, differing technologies were available to provide access and enable utilization of forest resources at different points in time in different locations. Changing industrial technologies could alter wood utilization dramatically. So as economic relationships between markets and suppliers raised and lowered wood's value in particular locations,

 Methods and Approaches in Forest History (eds M. Agnoletti and S. Anderson)

resulting utilization intensities varied considerably even in the same timeframe. As long as wood was society's primary raw material, wherever on the American frontier and in the settled regions behind it that humans encountered forests, supply–demand principles operated as they had for millennia. But subsequent forest conditions might vary considerably depending on when and how an area was first settled.

Determining forest history then becomes a site-specific concern, and differences in sites mean that site-specific studies may require and benefit from a multiplicity of analytical techniques and tap a variety of information sources. Thus one purpose of this chapter is to illustrate how research methods can be integrated in environmental history studies. To do so, the chapter will focus on how these studies illustrate differences caused by: (i) time of settlement; (ii) access to market economies and wood demands; (iii) scale of investigation; (iv) techniques for analysis related to study aims; and (v) the amount of funding support for the research.

This chapter examines parallels and differences between two forest history studies. Both studies involve specific forest sites. Spatial scales of the investigations differ, but the temporal frameworks are nearly the same. One investigation examines land use and environmental history of a Piedmont tract in Wake County, North Carolina, called the Harris Research Tract (HRT) leased by North Carolina State University from Carolina Power and Light (CP&L). The other study examines ecosystem changes in the central Appalachian mountains, surrounding the Red Hill Lot (RHL) or Military Lot 528, in Maryland's Garrett County. Overall aims of the historical investigations are the same – to understand present forest conditions as they reflect influences from the past. However, specific project objectives and therefore techniques for information gathering and analysis differ.

2 Geography of the two sites

North to south, approximately 460 km separate the sites, which are situated at about the same longitude. However, the HRT is 192 km from the Atlantic seacoast and the RHL is about 480 km from the Atlantic, though it is only about 274 km from Maryland's Chesapeake Bay.

The HRT includes 496 ha situated on the Piedmont's southeastern edge at approximately 98 m elevation. From highest to lowest point the tract's elevation changes 35 m, with the ridgetops rounded and side slopes gentle. Mayodan sandy loam soils dominate the site's ridges, with alluvial soils along the stream drainages. The HRT occurs where the transitional long-leaf pine (*Pinus palustris*) forest type was originally dominant. This transitional community is a variation of the extensive long-leaf pine forests that once occupied an estimated 60–90 million acres of the southeastern coastal plain (Frost, 1990, 1993).

The RHL itself covers only 12.6 ha, although the lot is just the focal point of the larger study area. At approximately 823 m elevation, the RHL is within North America's mixed mesophytic forest community. Its elevation changes 49 m from the highest to lowest point, with gradients typical of the surrounding area. It was originally forested with white pine (*Pinus strobus*) and hemlock (*Tsuga canadensis*) in its low and relatively flat portion. Northern hardwoods occur on the drier slopes above.

Climate of the HRT is humid subtropical and, on average, the area receives 118 cm rainfall $year^{-1}$, with about 1 cm of annual snowfall. Large frontal systems sweep across the southeast region in all seasons, distributing rainfall fairly evenly throughout the year but spawning localized thunderstorms in summer. Large cyclonic storms from the Atlantic Ocean can occur in the late summer and autumn. The climate in Garrett County is humid continental with a rainfall of about 113 cm $year^{-1}$, but the snowfall averages approximately 178 cm $year^{-1}$ (Stone and Matthews, 1974).

3 Settlement history

These sites were originally settled at about the same time period, in the second wave of European movement inland about 150 years after the earliest colonists settled the tidewater areas. In both Carolina and Maryland, the earliest transportation and commerce depended almost exclusively on waterborne conveyance. Overland travel was difficult for some time due to roads that were virtually impassable when wet.

Penetration to the Carolina interior from the coast was especially difficult and therefore slow. Permanent settlement in the Carolina colony first occurred when settlers from Virginia moved into the Albemarle Sound area in the 1650s. North Carolina's first town was not settled until after 1700. Farther north, the Maryland colony was founded in 1632 when an expedition sailed up the Chesapeake Bay to a small island in the lower Potomac river. Soon afterward, this group moved to the mainland, and thereafter plantations developed in tidewater areas of the Potomac river and Chesapeake Bay.

Amerindian populations had abandoned colonial Maryland's territory by 1700. By 1730 native populations in the Carolina Piedmont had been reduced to perhaps no more than 2000 individuals (Bode, 1997). At neither the HRT site nor the RHL site has evidence of permanent Amerindian habitation been discovered.

By the early to mid-1700s, immigration to the North Carolina Piedmont was beginning. Title searches indicate that John Holleman probably acquired a portion of the HRT lands as early as 1809 (Bode, 1997). Dates of earliest ownership for other parcels that now make up the HRT are indefinite but probably also occurred in the early 1800s (Bode, 1997). The earliest settlers in what would become Garrett County also came in the 1750s. However, Britain's Privy Council proclaimed in 1763 that these lands were off limits to colonial settlement: 'no governor or commander in chief of our other colonies or plantations in America do presume for the present, and until our further pleasure be known, to grant warrants of survey or pass patents for any lands beyond the heads or sources of any of the rivers which fall into the Atlantic Ocean from the west or northwest' (Encyclopaedia Britannica, 1976, p. 85). Prompted by the desire to ease tensions with native populations and the French, the proclamation was generally ignored by colonists along the frontier. In fact, official surveys were commissioned as early as 1768, just 5 years after the Privy Council's proclamation (Brown, 1896).

Title transfers were traced to 1878 at the Garrett County courthouse, and then at the Allegany County courthouse in Cumberland for years prior to 1878. Deeds indicate that Military Lot 528 (the RHL) was originally part of the large group of 50-acre (20.25 ha) lots bought by Thomas Johnson, who at one point owned over 6075 ha in Garrett County. It is not clear, however, when a cabin of white pine logs was erected on the RHL. It may have been erected by a squatter or by an indentured tenant. Similarity of construction to that of other area cabins whose origin has been specifically dated suggests that it was built around 1770.

4 Transportation key to market economics and wood demands

Garrett County's position proximal to major transportation routes led to early and rapid timber exploitation. The National Road (now US 40) started from Cumberland, Maryland, 19 km to the east, and provided a route through Garrett County to lands opening up in the Ohio valley. Begun in 1811, the road was completed in 1818, after which an almost steady westward stream of settlers passed through the county. This route also provided access to eastward markets for white pine and hardwood lumber cut in the uplands. Cumberland became a major departure point, and the building of wooden wagons became an important local industry. Within two decades, railroads extended into Garrett County from Cumberland. These rail lines first followed the Potomac and Savage river valleys and, later, another route was cut northwestward across Big Savage mountain into the Casselman river watershed to the northwest. Access by rail coupled with use of steam sawmills meant timber could be reached and sawn at considerably faster rates than before. Farther east, Baltimore was becoming a major port, and its shipbuilders needed timber for hulls and masts. Development of the C&O canal from Georgetown, DC, to Cumberland along the Potomac river provided another mode for transporting lumber.

The HRT was also near two roads, but these routes did not lead to major timber markets. In the mid-19th century two stagecoach routes, the Haywood Road from Raleigh and the Durham to Fayetteville Road, intersected at the village of New Hill 2.4 km northwest of the HRT (Murray, 1985; Bode, 1997). As early as 1796, investors began to build a canal on the Cape Fear river, which flows just a few miles south of the HRT (Ross, 1965). However, this effort failed and river navigation to the coast was thwarted by the fall line, which proved unbreachable. Rail access near enough to the HRT area to have any impact did not develop until the 1870s.

5 Land use trends

Though both sites were near regionally important transportation routes, forest product utilization and resource extraction techniques were markedly different. To understand current conditions at each site, it is also useful to consider their specific histories in the context of regional development. Though agriculture and livestock raising were ubiquitous features of American rural life during this era, climate and topographic differences in these locations created different local land use patterns. Local market economics and regional economic development patterns affected woodlands differently.

Despite its modern-day appearance as a rural and rustic environment, the RHL lot and vicinity were dramatically affected by early industrial development just a few miles away. The RHL was surrounded by sawmills from the early days of settlement. Brown (1896) notes that 'the first saw mill in Garrett county was built and owned by Philip Hare about the year 1790 on Meadow Run . . .' (p. 54). Water-powered mills were subsequently built even closer to the RHL in 1815 and 1837. The early mills cut primarily white pine, and even the first steam sawmill, built in 1837, was built to exploit 'a splendid lot of 250 acres of pine' which was cut in about 3 years (p. 55). Two more steam mills were built along Little Savage river around 1840, one north and one south of the pike. Because white pine grew on the moister and more productive sites, its harvest would have usually been followed by agricultural clearing. Stumps would be pulled and typically burned so the fields could be ploughed. However, a curious regional custom was noted by Beachy (1929): in some instances white pine stumps pulled from the fields were used to create fencerows that remained intact decades after the trees' extraction.

Industrial development just a few miles away in the George's Creek valley of Allegany County drove the need for timber extraction in hardwood forests of eastern Garrett County. Mining, iron production, tanneries and fire-brick refractories were all concentrated in this valley and required large supplies of mine props, tanbark and fuelwood (Sudworth, 1900). Another principal timber use after the B&O railroad reached Cumberland in 1842 was for railroad ties. Replacement of untreated ties that deteriorated quickly provided a strong local market for hardwood logs even into the modern period.

Curran (1902) reports that Garrett County's forest area had been reduced to 54% by 1902. Of that area, virtually all of the valley timber and side slope timberland had been either entirely cleared or culled for merchantable material. All that remained intact was the highest ridge timber but this too was soon culled for mine props to supply burgeoning coal mines in the George's Creek valley. Changing industrial technologies could alter wood utilization dramatically. In western Maryland, bituminous and anthracite coal use to produce iron and steel decreased demand for charcoal. Mine props, for the coal mines and the fire-clay mines, became an important commodity, but the slash and residual material was left on the harvest site. Once dry, this wasted wood led to wildfires. Industrial decline in western Maryland early in the 20th century as the deep coal mines were depleted had profound effects on Garrett County's forests. Incentives for active forest management were not in place and the depletion due to high grading left many stands in degenerate condition.

The current bucolic condition of the HRT is more consistent with historic land use in the vicinity. The Piedmont of North Carolina was primarily an agrarian culture where industry was tied to crops and therefore less intensive in its fuel demands. By the mid-19th century nearly 90% of Wake County had been cleared and converted to pasture and agriculture. However, the percentage of the HRT itself that was ever cleared appears to have been considerably lower than the Wake County average (Bode, 1997). In fact, based on results of our investigation we believe well over half of the tract was never cleared, though all forested areas were cut at some point and some areas were probably cut more than once.

Forest stands on the Harris Tract today therefore represent a succession legacy from turpentining, hog grazing, partial logging and fire suppression (Bode, 1997). Bode's record of ownership and activity on the Harris Tract documents the dominant agricultural use in some areas as well as later timber extraction and turpentining. The frequent occurrence of pine fatwood stumps and occasional cat-faced trees provides reminders that a turpentine distillery operated in nearby New Hill, and a mill sawing long-leaf timber operated at Apex for some time.

Thus, we see that the earliest settlement in Garrett County only preceded industrialization by a few decades, whereas agriculture and animal husbandry dominated in Wake County. The advanced transportation systems that penetrated the western Maryland mountains early encouraged rapid resource extraction and depletion, followed by relatively unsustainable practices. Less industrial and commercial activity in southern Wake County allowed a remnant of original forest species to persist. However, a fundamental change in management practice, namely the suppression rather than active use of fire, still led to changes in forest composition. This change, prevalent across the south, affected the viability of the long-leaf ecosystem (Landers *et al.*, 1995).

6 Modern and current conditions and trends

From the early 1800s until 1949 much of the RHL was actively farmed, with orchard and pasture occurring on portions of the parcel. About half of the property remained in a periodically cut hardwood stand. A neighbour named McKenzie, for instance, selectively cut second-growth trees (high grading) for lumber and built a barn on the property in the mid-20th century (McKenzie, personal communication). But by the 1960s, soil on the cleared slope was worn out and, late in the decade with federal assistance funds, the owner had most of the open area planted to mixed conifers.

Garrett County today reflects the accumulated effects of public policy and development economics upon the mountain landscape. Regional economic development and resource extraction have continuously exploited some areas of the timber resource. But the condition of the forests also reflects a history of forest management and stewardship fostered by public agencies since the early 1900s. Government incentives for private landowners and planting programmes on state lands contributed to an increase in forested acreages. Because the climate in western Maryland is harsh enough that farming can often be a marginal proposition, many farmers gave up and their land has naturally reverted to forest.

The result of all these factors is that about 71% of Garrett County is now forested (Frieswyk and DiGiovanni, 1988). The State of Maryland owns and manages 14,000 ha of Garrett County forests. Much of this area is mature deciduous forest on steep slopes unlikely to be cut extensively in the foreseeable future. Though some timber may be harvested, the Savage River State Forest is mainly managed for watershed protection, recreation and gamelands or other wildlife habitat. Black bear and turkey populations, for example, have risen steadily.

Today, Wake County is part of one of the fastest developing areas in the United States – the Research Triangle. Though 47% of Wake County was still forested in 1990, rapid population growth since then is reducing that amount steadily. In contrast, Garrett County is experiencing modest growth. Vacation and second home building has only minimally reduced the amount of forest from 71% of Garrett County in 1986.

7 Scale of investigation

From the beginning it was obvious that these investigations would differ in several scalar respects. As a result, the finer grained analysis pursued concerning the HRT contrasts markedly with the broader regional scope of the Garrett County study. Some of these differences relate to study aims as an outgrowth of my associations with the RHL and HRT sites. But proximity of the sites to my campus office (460 km vs. 32 km) and funding for the studies ($0 vs. $59,000) differ substantially as well. In a sense, current land management considerations generated both historical investigations, but they have evolved very differently.

The Garrett County study originated from my friendship with Morris Willey, current owner of the RHL and its 18th century cabin, and because of my family roots in western Maryland. My association with the HRT is much shorter (since 1987). It results from teaching a forest planning class for which the HRT serves as our laboratory subject. The RHL study grew from questions about management plans written for the RHL since the mid-1960s. These questions created a desire to understand how past practices led to conditions documented in the first plan. The lack of management plans for the HRT became primary motivation for our research there. As was noted above, the HRT is leased by CP&L to the North Carolina State. That relationship and our prior use for teaching enabled me to secure research support from CP&L. The grant paid for 2 years of graduate students' support and establishing field studies. As we proceed with management and a restoration project, further support for monitoring and ongoing research are expected.

The original temporal framework for the RHL study was going to be the last 40 years, a bit longer than the time span covered by forest management plans written for the tract. In contrast, the HRT study began with expectations that we would examine the period from the earliest settlement (mid to late 1700s) in order to craft plans for the future. In actuality, as shown above, the time periods encompassed in the studies are virtually the same.

As was noted earlier, the RHL is only 12.6 ha while the HRT is 496 ha. But the RHL became a focal point for the larger regional investigation, and that region is considerably larger than the HRT itself. Bode's (1997) study treats the regional history of Piedmont, North Carolina, in general and the specifics of the turpentine industry and commercial logging interests in the vicinity. Most notably, however, Bode's work focuses on agrarian impacts and related uses on the 496 ha tract, an emphasis resulting from our particular interest in the long-leaf ecosystem and the site's restoration potential. Our subsequent work will focus exclusively on the HRT and its ongoing management and restoration history.

8 Techniques for data gathering and analysis

Scott Bode's history of the HRT was a traditional archival search supplemented by an ethnographic investigation using personal interview techniques.

The aim was to tie the broad historical view of the region discernible through documentary sources to the actual landscape conditions occurring on the HRT over time.

Subsequent to Bode's study, Douglas Parker initiated botanical studies on the tract in 1997. His goal was to determine the character of existing mixed pine–hardwood stands and their relation, if any, to the presumed character of the Piedmont transitional long-leaf forests reported in the literature. Parker's botanical study used the North Carolina Vegetative Sampling Protocol (Peet *et al.*, 1996). It therefore provided the details and data required to document a site for North Carolina Natural Heritage Program records. Now complete, this work will lead to the restoration of a long-leaf site.

Materials gathered about Garrett County and the RHL are primarily documentary because to date I have lacked research funding for site-specific studies. Abundant narrative information and statistical data concerning the history of Garrett County and its forests are available. But few specific details about the RHL beyond its land tenure chronology as indicated by title transfers have been gathered. To date, no detailed forest community or botanical study has been accomplished, and particular land use practices and their probable impacts have not been investigated. Such a study may be initiated in the future.

9 Conclusion

On the basis of the work accomplished for the HRT project, we have documented that a significant portion of the tract (at least 182 ha) should be included in the North Carolina Natural Heritage database as an example of the Piedmont transitional long-leaf community. We also have established that long-leaf restoration efforts will be initiated as part of the plan for managing the HRT. The restored area can serve as a research and demonstration location unique in Wake County and may be the largest example of the transitional long-leaf community in North Carolina. In this instance, a historical anthropological investigation prompted a detailed botanical study, and the combination of methods will inform a restoration project. That restoration project will be incorporated into adaptive management of a unique ecosystem component.

Study of the RHL, while less specific than the HRT study, similarly underscores the importance of understanding historical access to broader market economies and the wood demands inherent in these markets. Transportation systems and technologies capable of handling bulky wood products and conveying them to markets instrumentally raised and lowered wood's value in particular locations. That value determined the manner in which forests were treated.

In both cases, transportation and access have continuously affected the relationship between the forest resource and its users. However, these studies demonstrate how important differences in timing of utilization may be to the subsequent condition of forests. Recognizing the impact of the past on the future can entail detailed analyses of current conditions; thus a range of technical analytical skills may be useful to explain and articulate the implications of broad historical studies.

References

Beachy, L.J. (1926) From 'forest primeval' to wheatfields. *Motor Travel* February, 23–25.

Bode, S.M. (1997) Land use and environmental history of the Shearon Harris Tract. MS thesis, North Carolina State University, Raleigh.

Brown, J. (1896) *Brown's Miscellaneous Writings Upon a Great Variety of Subjects.* J.J. Miller, Cumberland, Maryland.

Curran, H.M. (1902) *The Forests of Garrett County.* The Johns Hopkins University Press, Baltimore, Maryland.

Encyclopaedia Britannica (1976) *The Annals of America*, Vol. 2. Encyclopaedia Britannica, Chicago.

Frieswyk, T.S. and DiGiovanni, D.M. (1988) *Forest Statistics for Maryland – 1976 and 1986.* Forest Experiment Station Research Bulletin NE 107. USDA Forest Service NE.

Frost, C.C. (1990) Natural diversity and status of longleaf pine communities. In: Youngblood, G. and Frederick, D.L. (eds) *Forestry in the 1990's – a Changing Environment.* Society of American Foresters, Bethesda, Maryland.

Frost, C.C. (1993) Four centuries of changing landscape patterns in the longleaf pine ecosystem. In: *Proceedings of the Tall Timbers Fire Ecology Conference No. 18.* Tall Timbers, Tallahassee, Florida.

Landers, J.L., Van Lear, D.H. and Boyer, W.D. (1995) The longleaf pine forests of the southeast: requiem or renaissance? *Journal of Forestry* 93, 39–44.

Murray, E.R. (1985) *Wake: Capitol County of North Carolina.* Capitol County Publishing Company, Raleigh, North Carolina.

Peet, R.K., Wentworth, T.R., Duncan, R. and White, P.S. (1996) *The North Carolina Vegetation Survey Protocol: a Flexible, Multipurpose Method for Recording Vegetation Composition and Structure.* University of North Carolina, Chapel Hill.

Ross, M. (1965) *The Cape Fear.* Holt, Rinehart and Winston, New York.

Stone, K.M. and Matthews, E.D. (1974) *Soil Survey of Garrett County, Maryland.* USDA Soil Cons. Serv. and Maryland Ag. Exprmt. Stn.

Sudworth, G.B. (1900) *The Forests of Allegany County.* The Johns Hopkins University Press, Baltimore, Maryland.

Index

Abies 236
  alba 109, 150
    in construction 235, 237, 238
  cilicica 236
  nebrodensis 236
  pinsapo 236
Acer 85–86, 88
  platanoides 69
  pseudoplatanus 141, 149, 150–151, 152
African forests 10, 43–44
  Amani Nature Reserve 180
  East Usambara Rain Forest Nature Reserve 180–181
  Eastern Arc Mountains 173–187
  Kissidougou 43, 175
  Rift Valley escarpment 176
agriculture affecting landscape 174, 178–179, 176–219, 235
Aleppo pine *see Pinus halepensis*
alnoculture 10, 189–202
  methods 191–196
  woodland composition 191
Alnus 141
  as Neolithic building material 234
  glutinosa 69
  incana 191–196, 197–198, 201
Alpine valleys forestry studies 139–155
  land use development 141–142
  landscape system changes 149–151
  methodology 142–146
  population migration 147–148
  study site 140–141
  survey summary 151–153
  written history sources 146
Amani Nature Reserve 180
American forests 8–10
  Adirondack Park 26
  Blue Mountains 41, 44
  Harris Research Tract, N. Carolina 266–274
  Red Hill Lot, Garrat County 266–274
  White Mountains 242–245
  Yellowstone National Park 167
American restoration concepts 165–172
American Society of Environmental History (ASEH) 22, 25
ancient forest 38, 42, 49, 168
anthracology 15
  and charcoal burning 80–91
anthropology 11
  and dendrochronology 243
  as forest research tool 175–176, 203–214
  cognitive 204

anthropology *continued*
images of fertility 209
sacred aspect of trees 205, 206, 234, 235
Tuscan woodland and the community 206–208
Veneto region woodland and the community 208–210
Araucaria imbricata 26, 27
archaeobotany
and charcoal 81
as forest history research tool 221–230
archaeogenetics 11
archaeology and dendrochronology 243
ash *see Fraxinus ornus*
Atlas cedar *see Cedrus atlantica*

Balée, W 5
Bamford, PW 3
beech *see Fagus sylvatica*
Betula 158, 217, 218, 219
pendula 69, 141, 151
and coppicing 109–110
pubescens 69
Bialowieza 258–259
biodiversity 10, 53, 55
and forest management 251–263
and GIS surveys 162–163, 258–261
biome studies 256–259
Bruntland Report (1987) 51
cartographic document use 260–261
Club of Rome report (1972) 51
Convention on Biodiversity (1992) 253
definition 252–253, 256
estimates of species totals 255
history 253–255
methodology for research 256–262
reduction 38, 40, 41
Rio Conference and Agenda 21 (1992) 51, 253
stational rather than specific 256, 257
Stockholm Earth Summit (1972) 51
biogeography 253
biological archives 125–126
birch forests 159, 217
Black Forest, Baden 107–218
agricultural pattern 109, 213–214
climate 210–211
coppice varieties 209–210
forest policy 216–217
forest structure 211–212
forest use 214–215
Blue Mountains 41, 44
boat building 234
Bonnier, Gaston 255
boreal forest landscape changes 157–163
Bormann, FH 5
bristlecone pines *see Pinus longaeva*
Bruntland Report (1987) 51

Calabrian pine *see Pinus brutia*
carob *see Ceratonia siliqua*
Carpinus betulus 84–88, 217, 218
and coppicing 109
Castanea sativa 85–86, 88
and coppicing 109–110
fruit use replaced by potato 114
fruit value 88
woodlands defined as fields 190
Cedrus
atlantica 237
in pyramids 235
Ceratonia siliqua 237
charcoal burners and woodmen 207
charcoal burning 15, 31, 32, 42, 79–91, 116, 123, 126
byproducts 81, 88
experimental kilns 89–90
French sites 81
Ligurian Apennine sites 82–84
taxa identification 84–88
chestnut *see Castanea sativa*
climate change 96, 139
climatic events 234, 238–239, 243–244, 245
Club of Rome report (1972) 51
comparative studies 15–17, 24–26, 29–33
data comparison 30
economics and demography 31
law and politics 31
micro and macrolevel history 32
social studies 31
see also forest history
Conference for Forest Protection in Europe, 1993 61
coniferous forests 61, 63–64
see also tree species

conservation 37–38, 39
landscape resources 10
construction and timber use 216, 234, 235–238
coppicing 15, 63, 107–118, 122
alder treatment 191, 197–198
byproducts 109, 114–115, 116, 117
species harvested 109–110
Cornus 85
Corylus avellana 84–85, 86–87, 150
and coppicing 109–110
soil 72
stool analysis 67–77
altitude 68
growth and age 73–76
methodology 68–69
study area 69
surrounding vegetation 69, 71–72
Crataegus monogyna 192–194
cult of trees 204, 205, 234
Cupressus sempervirens 235

Dauerwald 50
Deevey, S 5
deforestation 42, 235
subjective interpretations 175
through industrial demand 62–63
degeneration 168–171
degradation 168–171, 174, 179, 185, 208
and population repression 43–44
concept definition 182–183
demo-ethno-anthropological cultural heritage 204
demographic influences 31
demology 204, 206
dendrochronology 14, 234, 236, 237, 242–244
Di Berenger, Adolfo
Dell'antica storia e giurisprudenza forestale in Italia 2
Douglas fir *see Pseudotsuga menziesii*

East Usambara Rain Forest Nature Reserve 180–181
Eastern Arc Mountains
Iron Age communities 178–179
landscape change research 176–181
microclimate 178
Eastern hemlock *see Tsuga canadensis*
ecological awareness 61–65
ecology 7
colonial interpretation 164–175
and economic resource management 95
human or cultural 11
and population dynamics 94, 98
self-regulation 36
economic history 7–8, 31
and forestry 81, 93–105
and tree products 234
archives and forest history 122–126
economics history 37–38
and forestry 6
ecosystem adaptation 5, 183
ecosystem management 41
elm *see Ulmus campestris*
enclosures 97
endangered species and timber production 122
environmental damage and restoration 165–172
environmental history 21–28
and forests 93–105
research and historical geography 131–134
environmental law 26
Euonymus europaeus 84–85
European forests 129–138, 216–219
and xylolography 233–239
European larch *see Larix decidua*
expansion of forests 9, 43

Fagus sylvatica 11, 84–86, 109, 150–171, 191–196, 217–219, 258
in construction 237
Fairhead J and Leach M
Misreading the African Landscape 43, 175
farming system analysis (FSA) 142
fertilizer 62, 99
Finnish forests 67–76
fire and vegetation changes 158–159, 161–162
fire damage – evil or benign 171
fire management 37, 43, 120, 121, 196, 199–201
Forest History Institute, Freiburg 3
forest definitions
autopoietic 53–56
classic 53–56
economic resource 48–50, 54–55, 59–66

forest definitions *continued*
- new forestry 39–41
- planetary resource 51
- romantic 60–61
- superorganism 50–51

forest history 11–15
- 'long series' lack 12–13
- aerial photo use 175
- affected by local politics 180–181
- anthropology 203–214
- archaeobotanical data 221–230
- archaeological 185
- biological archives 9, 14–15
- data processing 132
- dendrochronology 14, 234, 236, 242–244
- forest surveys 125, 160–162, 162–163
- genetic data 223, 225–227
- geographic analysis 131–134
- interpretation 35–46
- laws 31
- oral 13–14, 180, 184
- origins 1–4
- pollen analysis 14–15, 80, 185, 215–216, 217
- record keeping 12–13
- satellite photo use 175
- scientific sources 184
- in Sweden 122–126
- written sources 12–13, 30, 100–101, 146–147, 180, 219
- *see also* comparative studies; landscape history

Forest History Society, USA 3–4, 8, 20–28
- awards 23
- interdisciplinary research 25–26
- journals and publications 21–22, 23, 24–25, 39
- library and resources 22–23
- manifesto 24
- oral history programme 23

forest history theory 39–40

forest law 30, 31, 44, 65, 98, 122
- Consegne dei Boschi 190–196
- and coppicing management 116
- as evidence for history 124
- and sustainable yield 122

forest management 158, 160
- adaptive 183
- economic 96–97
- positivism or complexity 47–58

forest products 99, 196, 207, 216–217, 234, 269–270
- barrel hoops 69
- leaf hay 150–151, 152, 216–217
- potash 30–36, 120–121, 123
- tar 121, 123, 126
- *see also* charcoal burning; timber production

Forest Products History Foundation *see* Forest History Society, USA

forest records
- combined with biological archives 125–126

Forest Stewardship Council certification 98

forestry paradigms 35–44

forestry science development 48–49, 56

forestry systems
- classic or autopoietic 53–56

Forstgeschichte 3

Fraxinus
- *excelsior* 69, 141, 149, 150–151, 152, 217
- *ornus* 84–86, 88

Frazer, James
- *The Golden Bough* 204–205

French forests 81
- Lure mountain 257–258
- Saint-Gobain forest 256–257
- Vosges mountains, Alsace 107–118

fuel 234

genetics 11
- as forest history research tool 221–230

geochronodiversity 253, 261

geographic information systems (GIS) 16–17, 131, 258–261
- applied to boreal forest landscape 157–163

geometrics 259–262

German forests 2–3, 9
- Black Forest, Baden 107–118
- Westphalia 60–64

grazing 120, 148, 150, 151, 153, 199, 201, 216, 270

grey alder *see Alnus incana*

Group d'Histoire des Forêts Francaises 14

Harris Research Tract 266–274
- current forestation 271

data collection and analysis 272–274
geography and climate 266–267
land use 270–271
settlement history 277
study framework 272
transport links 269
hazel *see Corylus avellana* L.
herbicides for species management 122
historical ecology 5–6, 9–10, 16
historical techniques 12–17
field investigation 17
multiple analyses 17
historical–geographical forest study 9, 129–138
hop hornbeam *see Ostrya carpinifolia*
hornbeam *see Carpinus betulus*
human role in forest history 4, 5, 7, 9, 36, 40, 42–43, 60, 94–95
hunting and shooting 201, 208

industrial silviculture 7, 37–38, 39
industrial timber use 62–63, 98, 99, 116, 120–121, 235, 269–270
German iron industry 60
insect infestation 37, 41
International Union of Forestry Research Organization (IUFRO) i–ii, 4, 5–6, 17
Iron Age communities 178–179, 218
isozyme markers 15, 223
Italian cyprus *see Cupressus sempervirens*
Italian Forest School, Vallembrosa 2
Italian forests
Abruzzo National Park 167
Ligurian Appenines 10, 79–91, 82–84, 189–202
Mastallone Alpine valley region 9, 139–155
Orgia, Tuscany 206–202
southern Italy 221–229
Treviso wetlands, Veneto 208–210
Italian restoration concepts 165–172

Juniperus 236
in pyramids 235

Laburnum 141, 151
land use
classification schemes 142–143
land utilization system (LUS) 142, 150–151, 153
land utilization type (LUT) 142–143
landscape change 99–100
aerial photo interpretation 143–144
in Alpine valleys 139–155
diagnostic methods 143–146
field surveys 144–146
pattern analysis 144
vegetation as diagnostic criteria 143
landscape history
aerial photography 185
archaeological sources 185
archival resources 185
in Eastern Arc Mountains 173–187
oral sources 184
pollen analysis 185
research project definitions 186
scientific sources 184
see also forest history
landscape transformation concepts 181–182
Larix decidua 141, 149
in construction 235, 237
Latin American forests 42
Le Guyader, Hervé 255
Leach, M *see* Fairhead J and Leach M
Ligurian Apennines forests 189–202
Likens, GE 5
Linné, Carl von
Systemae Naturae 158, 255
lodgepole pine *see Pinus contorta*
logging industry *see* timber production
Lure mountain 257–258

Malus sylvestris 69
management systems
adaptive 53, 55
economic 48–50
monoculture 49
systemic 55–56
metalworking 234
micro and macrolevel history 32
monkey-puzzle tree *see Araucaria imbricata*

nature and culture 11, 37, 41, 60–61, 203–214, 242, 245–246
Normalbaum 48
North Carolina Natural Heritage Program 273
Norway spruce *see Picea abies*

Odum, E 5
organicism 50
Ostrya carpinifolia 84–86, 88

palaeobotany 11
and charcoal burning 80–91
palaeoecology 14–15
pasture 63
patch dynamics 149, 153, 182
Perona, V 2
Picea 217
abies 116, 120, 149, 158, 159
in construction 235, 236, 237, 238
pine forests 61, 63
Pinus
brutia 222–223, 224–225, 237
contorta 160
halepensis 11, 221–230
genetics 223, 225–226
leucodermis
in construction 236
longaeva 241–249
and timberline studies 244–245
dendrochronology 242–244
negra
in construction 236, 237
pinaster 224
pinea 224
ponderosa 41
sylvestris 50, 116, 120, 122, 158, 159
in construction 235, 238
Piussi, P 5
Pliny the Elder
Natural History 254–255
Polish forests
Bialowieza 258–259
political control of forests 31, 44
pollen analysis 14–15, 185, 215–216, 217
in charcoal 80
population
and ecology 94–95, 97
migration 97, 147–148
population change
in Eastern Arc mountains 176, 178–179
and *Pinus halepensis* introduction 228
Populus
in construction 237
tremula 141, 151, 158
potatoes replacing chestnuts 114
prehistoric culture and forests 215–220
process concepts 182–183
degradation 182
persistence 183
predictability 183
resilience 183
process dynamics 36, 158–159, 161–162
and prehistoric settlements 216–219
professional historians 29–33
scientists or humanists 6, 25
training 12, 13
Prunus spinosa 192–194
Pseudotsuga menziesii 237

Quercus
as Neolithic building material 235
calliprinos 237
cerris 191–196
deciduous 85
in construction 216, 236, 237, 238
petraea 117, 150–151
and coppicing 109–110
pubescens 117
robur 109, 141

Rackham, Oliver
Trees and Woodland in the British Landscape 4–5,11
radar 17
radiocarbon dating 243
Red Hill Log, Garrat County
current forestation 271
data collection and analysis 273
geography and climate 267
land use 269–271
settlement history 268
study framework 272
transport links 268
reduction of forests 7, 9, 43–44
regional comparative studies 173–187
resource management 41, 42, 43, 48–49, 94–95, 96–97, 124–125
adaptive 183
restoration concepts 165–172
damage by people or nature 170–171
degradation or degeneration 167–169
landscape as garden or wilderness 166–167

Reuss, E 3
Rhododendron 141
Rio Conference and Agenda 21 (1992)
 51, 253
river valley studies 100–103
road construction affecting forests
 209–210, 268–269, 273
Robinia pseudoacacia
 and coppicing 109–110
romantics 60–61
Rosaceae 85
Rubner, H 4
Rubus
 hirtus 192–194
 idaeus 192–194
 ulmifolius 192–194

satellite imagery 17, 258–261
savanna-forest landscape 43
Scots pine *see Pinus sylvestris*
Sequoia 242
Sereni, E 3
sharecroppers 207
shelterwood felling 49
silver fir *see Abies alba*
silviculture
 archives used for forest history
 132–133
 financial 48–49
 systemic 51, 55
social history and forests 6, 7–8, 31
socio-economic interpretation of forests
 132
Sorbus 85, 88, 141
 aucuparia 69, 151
spruce forests 50, 61, 63
 see also tree species
standardization of trees 48
Stockholm Earth Summit (1972) 51
sustainable yield 30, 36, 37, 38, 39, 41,
 48–49, 51, 53, 54–38, 158
 in Swedish boreal forests 122, 124
Swedish forests 9, 98, 100, 101
 boreal 119–128, 157–163

Tamaryx in pyramids 235
Taxus baccata 236
Theophrastus
 A History of Plants 254
Tilia 141
timber production 4, 8–9, 37, 55, 60,
 62–66, 121–122, 123, 160,
 161, 269–270
 and ecological awareness 7
 and production law 98
 archives 122–124
timber shortages 62–63, 65, 99
timber transport 217, 234
 in ancient Rome 235
 to prehistoric settlements 217
 by river 100–103, 121–122
 by road and rail 102–103
 and written records 219
timberline studies 244–245
transnational analysis 27
tropical forests 42–44
Tsuga canadensis 237
Turkey oak *see Quercus cerris*
Tuscan woodland 206–208

Ulmaceae
 decline through pollarding 217
Ulnus 217
 campestris 84–86, 88
 glabra 69, 141, 258

vectorial fusion 259–262
vegetation change 15, 143
 and fire adaptation 120
 in Rift Valley 176
 problems of interpretation 175
Viburnum lantana 84–86
Vogt, K 5
Von Hornstein
 Wald und Mensch 3
Vosges mountains, Alsace 107–118
 agricultural pattern 109, 113–114
 climate 110–111
 coppice varieties 109–110
 forest policy 115–117
 forest structure 111–112
 forest use 114–115

Waldgeschichte 3
wetlands 209–210
wildlife 101
 fire adaptation 120
wood remains preservation 80

xylology and forest history 233–240

yew *see Taxus baccata*

Zanzi-Sulli, A 5
Zurich–Montpellier floristic diagnosis
 144–146